OXFORD LOGIC GUIDES: 47

General Editors

ANGUS MACINTYRE
DOV M. GABBAY
DANA SCOTT

OXFORD LOGIC GUIDES

For a full list of titles please visit
http://www.oup.co.uk/academic/science/maths/series/OLG/

Set Theory

Boolean-valued Models and Independence Proofs

Third Edition

JOHN L. BELL

Professor of Philosophy
University of Western Ontario

CLARENDON PRESS • OXFORD

2005

OXFORD
UNIVERSITY PRESS

Great Clarendon Street, Oxford, OX2 6DP,
United Kingdom

Oxford University Press is a department of the University of Oxford.
It furthers the University's objective of excellence in research, scholarship,
and education by publishing worldwide. Oxford is a registered trade mark of
Oxford University Press in the UK and in certain other countries

Published in the United States of America by Oxford University Press
198 Madison Avenue, New York, NY 10016, United States of America

British Library Cataloguing in Publication Data
Data available

Library of Congress Cataloging in Publication Data
Data available

ISBN 978–0–19–960916–1

NOTE TO THIRD EDITION

In this new edition the background material has been enlarged to include an introduction to Heyting algebras, and chapters (also of an introductory nature) added on Boolean-valued analysis and Heyting-algebra valued models of intuitionistic set theory. In a new Appendix the basic concepts of category theory are outlined and Boolean and Heyting-algebra valued models presented from that standpoint, affording important insights into the nature of these models.

1 July 2004

J.L.B.

NOTE TO PAPERBACK EDITION

In this edition a few new references have been added, and some errors corrected. I am grateful to Dominic McCarty for providing a list of the latter.

13 August 2010

J.L.B.

PREFACE TO THE SECOND EDITION

The present book had its origin in lecture courses I gave at London University during the early seventies. In writing it, my objective has been to provide a systematic and adequately motivated account of the theory of Boolean-valued models, deriving along the way the central set-theoretic independence proofs in the particularly elegant form that the Boolean-valued approach enables them to assume. The book is primarily intended for readers who have mastered the material ordinarily dealt with in a first course on axiomatic set theory, including constructible sets and the Gödel relative consistency proofs. I have also assumed some acquaintance with mathematical logic, Boolean algebras, and the rudiments of general topology. In order to expand the scope of the book, and to develop the reader's skill in the subject, many problems (with hints for solution) have been included, some of a more sophisticated character.

The chief purpose in preparing this second edition has been to incorporate an account of iterated Boolean extensions and the consistency (and independence) of Souslin's hypothesis, material not included in the original edition. The addition of this new material (which appears in Chapter 6) necessitated that certain changes be made in earlier chapters: this requirement, together with the generous offer of the Oxford University Press to completely reset the book, afforded me the opportunity of substantially modifying the original text. The major changes in this regard include a new—and I believe more perspicuous—proof that the axiom of choice holds in the model, and a revamping of Chapter 3 on the independence of the axiom of choice in terms of group actions on the model. (I am grateful to Yoshindo Suzuki for suggesting the idea of bringing group actions into the foreground.) I have also grasped the opportunity of suppressing the references with which the original text was liberally peppered, and whose somewhat misleading nature proved vexatious to several reviewers. These references have now been replaced by a set of brief—but nevertheless, I hope, reasonably accurate—historical notes at the end of the book.

It will be evident that in writing a book of this kind I have incurred a heavy intellectual debt to the mathematicians whose work I have attempted—with some temerity, perhaps—to expound. In this respect I am particularly indebted to Dana Scott. His unpublished, but widely circulated, 1967 notes (Scott 1967) are familiar to all set-theorists as the *urtext* in the field of Boolean-valued models and their influence is to be observed throughout the book. Moreover, as Editor of the Oxford Logic Guides he offered much advice and assistance during the preparation of the original text and has generously provided a Foreword. It was at his suggestion that I embarked on the preparation of this new edition; I am grateful to him both for his encouragement and for his offer to have the text set at Carnegie-Mellon University using the formatting system TEX (which, as the reader will see, has resulted in a most elegant printed text).

It is also a pleasure to tender my thanks to John Truss for his careful reading of the manuscript of the original edition and his many ideas for improving it; Enrique Hernandez for his assistance in checking the typescript; Gordon Monro for his valuable comments on Chapter 6, which resulted in the eradication of many errors; Karin Minio and members of the Computer Science Department at Carnegie-Mellon University for their careful handling of the computer preparation of the text; and Mimi Bell and Buffy Fennelly for their expert typing of the manuscript. Finally, I would record my gratitude to Anthony Watkinson and the Oxford University Press, without whom the whole enterprise would never have been brought to fruition.

London
June 1984

PREFACE TO THE FIRST EDITION

The present book had its origin in lecture courses I gave at London University during 1972/3 and 1974/5. In writing it, my purpose has been to provide a systematic and adequately motivated account of the theory of Boolean-valued models, deriving along the way the basic set-theoretical independence proofs in the particuarly elegant form they naturally assume in this approach. The book is primarily intended for readers who have mastered the material ordinarily covered in a first course on axiomatic set theory, including constructible sets and the Gödel relative consistency proofs. I have also assumed some acquaintance with mathematical logic, Boolean algebras, and the rudiments of general topology. In order to give the book wider scope, and to develop the reader's skill in the subject, many problems (with hints for solution) have been included, some of a more sophisticated character.

It will be evident from the many references and attributions given in the text that I have incurred a heavy intellectual debt to the mathematicians whose work I have attempted—with some temerity, perhaps—to expound here. In this respect I am particularly indebted to Dana Scott. His unpublished, but widely circulated, 1967 notes (Scott 1967) are familiar to all workers in the field of Boolean-valued models and their influence is to be observed throughout the book. Moreover, as Editor of the Oxford Logic Guides he offered much valuable advice and assistance during the preparation of the text and has generously provided a Foreword. I would also like to thank John Truss for reading the manuscript and making many helpful suggestions for improving it, Enrique Hernandez for his assistance in checking the typescript, and Buffy Fennelly for her expert typing.

Finally, a cautionary word on references to the Bibliography. In a subject which has developed as rapidly as the one treated in this book, it is inevitable that many results and notions never reach the stage of official publication in their original form, but are instead quickly absorbed into that nebulous domain commonly known among mathematicians as 'folklore'. (See Dana Scott's Foreword for an illuminating discussion of the background and growth of the subject.) Accordingly, the reader should be warned that the bibliographical references in the text are *not necessarily* to original sources, but rather indicate the places from which I have drawn my material.

London   J.B.

June 1977

To Mimi

FOREWORD

Even if we were not required by Russell's Paradox to take care in formulating the axioms of set theory, we would nevertheless have many difficult questions to answer concerning infinite combinatorics and infinite cardinal numbers. Think for a moment of the Axiom of Choice, the Continuum Hypothesis, Souslin's Hypothesis, the questions of the existence of inaccessible and measurable cardinals, the problems of the Lebesgue measurability of projective sets, or of the determinateness of various kinds of infinite games. These are all questions of 'naïve' set theory, many involving little beyond the concept of the arbitrary set of real numbers. Perhaps it is only hindsight (helped on, to be sure, by Gödel's Incompleteness Theorem), but we would have been extraordinarily lucky if these intricate and often fundamental problems were capable of being settled *by logic alone*. By 'logic' here we mean first-order logic (the logic of connectives and quantifiers) together with some rather pure 'ontological' axioms of set existence (like the Comprehension Axiom). An axiom like the Extensionality Axiom, which says that sets are uniquely determined by their elements, is also sufficiently logical in character, because of its almost definitional nature. So we can throw it to the side of logic.

When we come to the Axiom of Choice, we begin to waver: it might be argued that it is implicit in the concept of the totally *arbitrary* set. On the other hand, there could be other notions of what it means to *determine* a set for which it would fail; thus, the act of assuming it is indeed axiomatic: it is 'self-evident' but not just a matter of logic. But then, perhaps it *is* a matter of logic after all, because the *finite* version is provable. In other words, first-order logic is strong enough for some conclusions, but it is in general too weak: we ought perhaps to allow 'infinitary' inferences also. And at this point we begin to wonder what is meant by logic. It would seem rather circular if in making set theory precise, we had to use set theory in order to make logic precise.

With regard to the Continuum Hypothesis most people, I feel, would call this assumption truly axiomatic, since it is so very special in *excluding* certain sets of reals of an intermediate cardinality. True, it can be stated in rather pure terms, and in *second-order* formulations of set theory it would be decided: only we cannot know which way. Thus, the argument for its logical character is rather thin. Certainly Gödel's consistency proof gives us no real evidence. Beautiful as they are, his so-called constructible sets are very special being almost *minimal* in satisfying formal axioms in a first-order language. They just do not capture the notion of set *in general* (and they were not meant to). The constructible universe is extremely interesting in itself (e.g. Jensen showed, among other things, that Souslin's Hypothesis *fails* if $V = L$), but there are very few who would want to assume $V = L$ once and for all. This leaves us more than ever with unsettled feelings as to *where* to draw the line between mathematics and logic.

The events in set theory since 1960 have in some ways made matters even worse. First there was the explosion in large cardinals beginning with the Hanf-Tarski discovery that the first inaccessible (already large enough for a cardinal) was *much* smaller than the first measurable. There was then a most elaborate development in the study of infinite combinatorics by a large number of researchers too numerous to mention here, but among whom Erdös has a central place. On the logical side these results had many model-theoretic consequences and in particular showed that the spectrum of stronger and stronger large-cardinal axioms was *very* finely divided. (The reader can refer to Drake (1974) for a survey.) And the work still goes on. Perhaps this sort of study is not basically disturbing, however, for it just shows—in nearly linear order—that if you want more you have to assume more. It did turn out that the existence of measurable cardinals was *inconsistent* with $V = L$, but so much the worse for the 'unnatural' constructible sets. And some comfort can be gained from the fact that any number of attempts at showing that measurable cardinals do not exist have failed even though much cleverness was expended.

It was in 1963 that we were hit by a real bomb, however, when Paul J. Cohen discovered his method of 'forcing', which started a long chain reaction of independence results stemming from his initial proof of the independence of the Continuum Hypothesis. Set theory could never be the same after Cohen, and there is simply no comparison whatsoever in the sophistication of our knowledge about models for set theory today as contrasted to the pre-Cohen era. One of the most striking consequences of his work is the realization of the extreme *relativity* of the notion of cardinal number. Gödel has shown that, by *cutting down* on the totality of sets, the cardinals (of the model L) would be very well-behaved. Cohen showed that by *expanding* the totality of sets the cardinals would be very *ill*-behaved. (Tiresomely difficult questions about the possible bad behaviour of *singular* cardinals from model to model is more than sufficient to make the point.) Of course we had realized that the cardinals of L might not be the cardinals of V (indeed, with *fewer* sets there are *more* cardinals, because there is less chance for a one–one correspondence), but we had no idea before Cohen (and those who so quickly jumped into the field after him) how much independence there could be. Thus we can make (if anyone would want to) $2^{\aleph_0} = \aleph_{17}$ and $2^{\aleph_1} = \aleph_{2001}$ in some model or the other, and even with these silly choices the size of $2^{\aleph_2}$ is not at all well determined (except that it has to be greater or equal to $\aleph_{2001}$ and has to avoid certain singular cardinals).

Cohen's achievement lies in being able to *expand* models (countable, standard models) by adding new sets in a very economical fashion: they more or less have only the properties they are *forced* to have by the axioms (or by the truths of the given model). I knew almost all the set-theoreticians of the day, and I think I can say that no one could have guessed that the proof would have gone in just this way. Model-theoretic methods had shown us how many *nonstandard* models there were; but Cohen, starting from very primitive first principles, found the way to keep the models *standard* (that is, with a well-ordered collection of ordinals). And moreover his method was very flexible in introducing lots and lots of models—indeed, too many models. Is it not just a bit embarrassing that the currently accepted axioms for set theory (which could be given—as far as they went—a perfectly natural motivation) simply did *not* determine the concept of infinite set even in the very important region of the continuum?

We should not get the idea that Cohen's method solves all problems. For example, Shoenfield's Absoluteness Lemma shows us why the 'simplest' nonconstructible set is Δ^1_3, and thus Cohen's models can only start their independence proofs at that level of the analytic hierarchy. Furthermore, we as yet have no exactly similar model-theoretic independence proofs *from* $V = L$, and this is certainly a very interesting problem.[1] Nevertheless, Cohen's ideas created so many proofs that he himself was convinced that the formalist position in foundations was the rational conclusion. I myself cannot agree, however. I see that there are any number of contradictory set theories, all extending the Zermelo–Fraenkel axioms: but the models are all just models of the first-order axioms, and first-order logic is weak. I still feel that it ought to be possible to have strong axioms, which would generate these types of models as submodels of the universe, but where the universe can be thought of as something absolute. Perhaps we would be pushed in the end to say that all sets are *countable* (and that the continuum is not even a set) when at last all cardinals are absolutely destroyed. But really pleasant axioms have not been produced by me or anyone else, and the suggestion remains speculation. A new idea (or point of view) is needed, and in the meantime all we can do is to study the great *variety* of models. It is the purpose of the present book to give an introduction to this study via the notion of Boolean-valued models. Chapter 2, however, ties up the approach with Cohen's original ideas, though avoiding the technicalities of the ramified languages as is usual in most later presentations.

The idea of using Boolean-valued models to describe forcing was discovered by Solovay in 1965. He was using Borel sets of positive measure as forcing conditions; the complications of seeing just what was true in his model led him, as I remember from a conversation at Stanford around September of that year, to summarize various calculations by saying that the combination of conditions forcing a statement added up to the 'value' of that statement (*cf.* Theorem 2.4 in

---

[1] Recently, however, Harvey Friedman in his paper 'On the necessary use of abstract set theory', *Advances in Mathematics*, vol. 41, 1981, pp. 209–280, showed that interesting combinatorial propositions about Borel functions are independent even from $V = L$. His methods use nonstandard models in a way he considers essential. It does not seem that the method can be called similar to Cohen's, however, and there seem to be good reasons for this difference.

this book and the surrounding discussion). Petr Vopěnka (1965) independently had much the same idea, but his initial presentation was brief and not so very attractive; so we were not much struck by his approach at first. In thinking over Solovay's suggestion, it occurred to me that by *starting* with Boolean-valued sets from the very beginning, many of the more tedious details of Cohen's original construction of the model were avoided. Solovay by November of 1965 had also come to this conclusion himself, and, as it turned out, this was what Vopěnka was actually doing. In the end, as was demonstrated by the paper of Shoenfield (1971) from the 1967 Set Theory Symposium, there is very little to choose between the methods: forcing and Boolean-valued models both come to the same thing. Psychologically, however, one attitude or the other may be more suggestive. Boolean-valued models *are* quite natural; but, when it comes to the proofs (and the *construction* of the right model), one often has to look very closely at the forcing conditions.

The whole history of the independence proofs is rather complex and it could only be made clear by going into the exact technical details. The two volumes of contributions to the 1967 Symposium (Scott 1971; Jech 1974) contain many of the original papers by Cohen and an historical paper on the Prague School by Petr Hajek. The lecture notes by Felgner (1971) also contain a very useful summary of basic results with many references to the original sources. In the present book, John Bell has made very good use of the lecture notes on Boolean-valued models, which were distributed at the Symposium, and which were prepared for typing by Kenneth Bowen and myself during the time of the conference from my handwritten manuscript. These are the notes (Scott 1967) mentioned in the bibliography; in writing them my main role was that of an expositor.

There are many references in the literature to the Scott–Solovay paper, which was to be published as an expanded version of the 1967 notes. This paper does not exist, and it is my own personal failing for not putting it together from the materials I had at hand. I discussed it several times with Robert Solovay, but we were not at the same institution and could not work very closely together. He drafted parts of certain sections, but he was working on so many papers at the same time that he did not have the opportunity to draft the whole paper. The present book essentially supplants the projected Scott–Solovay paper. Part of my own difficulty about writing the Scott–Solovay paper was the fantastic growth of the field and the speed with which it changed. During the winter of 1968–1969 I became profoundly discouraged because I felt unable to make any original contributions: any ideas I had were either wrong or already known. It is easy enough to say now that I should have been content to be a reporter and expositor, but, at the very moment when one is being left behind, things seem less pleasant. I put these remarks forward not as an excuse but simply as an explanation of why I could not complete what I set out to do.

Looking back on the development of logic and set theory it is very tempting to ask why the independence proofs were not discovered earlier. From the point of view of Boolean algebras we had the needed technical expertise in constructing complete Boolean algebras many years before Cohen. Perhaps model theory up to 1960 had concentrated too much on first-order theories. Actually in September of 1951, in a paper of Alonzo Church delivered at the Mexican Scientific Conference, a suggestion for Boolean-valued models of *type theory* had already been made.[2] However, the suggestion suffered from the fault that not enough care was taken over the Axiom of Extensionality, since Church recommends that at higher types one take *all* functions from one type to the other. Unless the equality relation and membership (application) is treated as on p. 23 of the present book, difficulties will arise. These difficulties would have been easily overcome, though, if anyone had tried to develop this clearly stated suggestion.

The 'first-order disease' is most plainly seen in the book by Rasiowa and Sikorski (1963). They had for a number of years considered Boolean-valued models of first-order sentences (as had Tarski, Mostowski, Halmos, and many others working on algebraic logic). Unfortunately they spent most of the time considering *logical validity* (truth in all models) rather than the construction of possibly interesting *particular* models. But even so, with all this machinery, no one thought to ask: how do we interpret second-order quantifiers? It would have been the most natural thing in the logical world, because the values of sentences with *arbitrary* Boolean-valued *relations* were already defined. The step from the arbitrary constant, to the variable, to the quantifier is obvious: it had already been taken at the first-order level. It was a real opportunity missed, and one missed for no good reason except the failure to ask the right question. And, if the question had been asked, the problems about cardinalities would have had to be faced. Well, there is no changing of history.

What about forcing? How new was this idea? As we have said, the application to set theory was strikingly new. Kleene, however, had already used a similar idea in recursion theory where, in studying degrees, he had to force a sequence of Σ_1^0-sentences. The wider model-theoretic significance was not appreciated, though, even if the technique was generalized in recursion theory. In studies of intuitionistic logic (both with Kripke models and with Beth models) the kinds of clauses similar to the forcing definition were quite well-known, but it did not seem to occur to anyone to employ intuitionistic logic in making *extensions* of models. Kreisel, in a paper at the Infinitistic Methods Conference in Warsaw in 1959, suggested briefly something very like forcing, but the plan lay quite undeveloped by him or any of his readers. However, after Cohen's original announcement, I pointed out the analogy with intuitionistic interpretations, and along these lines Cohen simplified his treatment of negation at my suggestion. Later the intuitionistic analogy was taken up more seriously by Grzegorczyk and worked out in detail in the book by Fitting (1969). What has transpired since that time is that the set theory in intuitionistic logic proper (not in the double-negative,

---

[2]The paper was published in English in *Boletin de la Sociedad Matematica Mexicana*, vol. 10, 1953, pp. 41–52, under the title 'Non-normal truth-tables for the prepositional calculus'. The remark Church attributes to Lagerström.

weak-forcing format) has become much more interesting, owing chiefly to the work in category theory by Lawvere and Tierney on topoi. Not only are there Heyting-valued models, but there are many more abstract 'sheaf' models. An introduction to models for intuitionistic set theory is presented by the author in Chapter 8. Up-to-date references and an extensive development of these ideas in a category-theoretic can be found in the two volumes of Johnstone (2002). (And we await a third volume as well.)

Oxford   Dana Scott
May 1977
(Revised, Pittsburgh, August 1984; Berkeley, August 2010)

CONTENTS

LIST OF PROBLEMS

0

BOOLEAN AND HEYTING ALGEBRAS:
THE ESSENTIALS

In this book we shall make considerable use of the theory of *Boolean algebras*. In the book's later sections we shall also employ the concept of *Heyting algebra*. We begin with a brief account of these notions. Fuller accounts include Balbes and Dwinger (1974), Bell and Machover (1977), Halmos (1963), Johnstone (1982), and Sikorski (1964).

Lattices

A *lattice* is a (nonempty) partially ordered set L with partial ordering $\leq$ in which each two-element subset $\{x, y\}$ has a supremum or *join*—denoted by $x \vee y$—and an infimum or *meet*—denoted by $x \wedge y$. A *top* (*bottom*) element of a lattice L is an element, denoted by $1_L(0_L)$ such that $x \leq 1_L(0_L \leq x)$ for all $x \in L$. When confusion is unlikely we drop the subscript and write simply 1 or 0. A lattice with top and bottom elements is called *bounded*. A lattice is *trivial* if it contains just one element, or equivalently, if in it $0 = 1$. A *sublattice* of a bounded lattice L is a subset of L containing 0 and 1 and closed under Ls meet and join operations.

It is easy to show that the following hold in any bounded lattice:

$$x \vee 0 = x, \qquad x \wedge 1 = x,$$

$$x \vee x = x, \qquad x \wedge x = x,$$

$$x \vee y = y \vee x, \qquad x \wedge y = y \wedge x,$$

$$x \vee (y \vee z) = (x \vee y) \vee z, \qquad x \wedge (y \wedge z) = (x \wedge y) \wedge z,$$

$$(x \vee y) \wedge y = y, \qquad (x \wedge y) \vee y = y.$$

Conversely, suppose that $(L, \vee, \wedge, 0, 1)$ is an algebraic structure, with $\vee, \wedge$ binary operations, in which the above equations hold, and define the relation $\leq$ on L by $x \leq y$ iff $x \vee y = y$. It is then easily shown that $(L, \leq)$ is a bounded lattice in which $\vee$ and $\wedge$ are, respectively, the join and meet operations, and 1 and 0 the top and bottom elements. This is the *equational characterization* of lattices.

Examples (i) Any linearly ordered set is a lattice; clearly in this case we have $x \wedge y = \min(x, y)$ and $x \vee y = \max(x, y)$.

(ii) For any set X, the power set PA is a lattice under the partial ordering of set inclusion. In this lattice $X \vee Y = X \cup Y$ and $X \wedge Y = X \cap Y$. A sublattice of a power set lattice is called a *lattice of sets*.

(iii) If X is a topological space, the families $O(X)$ and $C(X)$ of open sets and closed sets, respectively, in X each form a lattice under the partial ordering of set inclusion. In these lattices $\vee$ and $\wedge$ are the same as in Example (ii).

A lattice is said to be *distributive* if the following identities are satisfied:

$$x \wedge (y \vee z) = (x \wedge y) \vee (x \wedge z), \quad x \vee (y \wedge z) = (x \vee y) \wedge (x \vee z).$$

Interestingly, each of these conditions implies the other. For example, assuming the first condition, we have:

$$\begin{aligned}
(x \vee y) \wedge (x \vee z) &= [x \wedge (x \vee z)] \vee [y \wedge (x \vee z)] \\
&= x \vee [(y \wedge x) \vee (y \wedge z)] \\
&= [x \vee (y \wedge x)] \vee (y \wedge z) \\
&= x \vee (y \wedge z).
\end{aligned}$$

The converse is proved similarly.

In the sequel by the term 'distributive lattice' we shall understand 'bounded distributive lattice'.

An easy inductive argument shows that any nonempty finite subset $\{x_1, \ldots, x_n\}$ of a lattice has a supremum, or join, and an infimum, or meet: these are denoted respectively by $x_1 \vee \cdots \vee x_n$, $x_1 \wedge \cdots \wedge x_n$. An arbitrary subset of a lattice need not have an infimum or a supremum: for example, the set of even integers in the totally ordered lattice of integers has neither. If a subset X of a given lattice does possess an infimum, or meet, it is denoted by $\bigwedge X$; if the subset possesses a supremum, or join, it is denoted by $\bigvee X$. When X is given in the form of an indexed set $\{x_i : i \in I\}$, its join and meet, if they exist, are written respectively $\bigvee_{i \in I} x_i$ and $\bigwedge_{i \in I} x_i$.

A lattice is *complete* if every subset has an infimum and a supremum. The meet and join of the empty subset of a complete lattice are, respectively, its top and bottom elements. It is a curious fact that, for a lattice to be complete, it suffices that every subset have a supremum, or every subset an infimum. For the supremum (infimum), if it exists, of the set of lower (upper)[1] bounds of a given subset X is easily seen to be the infimum (supremum) of X.

---

[1] Here by a *lower (upper) bound* of a subset X of a partially ordered set P we mean an element $a \in P$ for which $a \leq x (x \leq a)$ for every $x \in X$.

Examples (i) The power set lattice PA of a set A is a complete lattice in which joins and meets coincide with set-theoretic unions and intersections respectively.

(ii) The lattices $O(X)$ and $C(X)$ of open sets and closed sets of a topological space are both complete. In $O(X)$ the join and meet of a subfamily $\{U_i : i \in I\}$ are given by

$$\bigvee_{i \in I} U_i = \bigcup_{i \in I} U_i \qquad \bigwedge_{i \in I} U_i = \overset{\circ}{\bigcap_{i \in I} U_i}.$$

In $C(X)$ the join and meet of a subfamily $\{A_i : i \in I\}$ are given by

$$\bigvee_{i \in I} A_i = \overline{\bigcup_{i \in I} A_i} \qquad \bigwedge_{i \in I} A_i = \bigcap_{i \in I} A_i.$$

Here $\overset{\circ}{A}$ and $\overline{A}$ denote the interior and closure, respectively, of a subset A of a topological space.

Heyting and Boolean algebras

A *Heyting algebra* is a bounded lattice $(H, \leq)$ such that, for any pair of elements $x, y \in H$, the set of $z \in H$ satisfying $z \wedge x \leq y$ has a *largest* element. This element, which is uniquely determined by x and y, is denoted by $x \Rightarrow y$: thus $x \Rightarrow y$ is characterized by the following condition: for all $z \in H$,

$$z \leq x \Rightarrow y \text{ if and only if } z \wedge x \leq y.$$

The binary operation on a Heyting algebra, which sends each pair of elements x, y to the element $x \Rightarrow y$ is called *implication*; the operation, which sends each element x to the element $x^* = x \Rightarrow 0$ is called *pseudocomplementation*. We also define the operation $\Leftrightarrow$ of *equivalence* by $x \Leftrightarrow y = (x \Rightarrow y) \wedge (y \Rightarrow x)$. These operations are easily shown to satisfy:

$$x \Rightarrow (y \Rightarrow z) = (x \wedge y) \Rightarrow z, \quad x \Rightarrow y = 1 \leftrightarrow x \leq y, \quad x \Leftrightarrow y = 1 \leftrightarrow x = y,$$

$$y \leq z \rightarrow (x \Rightarrow y) \leq (x \Rightarrow z), \quad x \wedge (x \Rightarrow y) \leq y$$

$$y \leq x^* \leftrightarrow y \wedge x = 0 \leftrightarrow x \leq y^*, \quad x \leq x^{**}, \quad x^{***} = x^*, \quad (x \vee y)^* = x^* \wedge y^*.$$

Any Heyting algebra is a distributive lattice. To see this, calculate as follows for arbitrary elements x, y, z, u:

$$x \wedge (y \vee z) \leq u \leftrightarrow y \vee z \leq x \Rightarrow u$$
$$\leftrightarrow y \leq x \Rightarrow u \text{ and } z \leq x \Rightarrow u$$
$$\leftrightarrow x \wedge y \leq u \text{ and } x \wedge z \leq u$$
$$\leftrightarrow (x \vee y) \wedge (x \vee z) \leq u.$$

Heyting algebras can also be characterized *equationally*. In fact we have the

Proposition 0.1 *Let L be a bounded lattice, $\Rightarrow$ a binary operation on L. Then $\Rightarrow$ makes L into a Heyting algebra iff the equations*

$$\text{(i) } x \Rightarrow x = 1 \quad \text{(ii) } x \wedge (x \Rightarrow y) = x \wedge y \quad \text{(iii) } y \wedge (x \Rightarrow y) = y$$
$$\text{(iv) } x \Rightarrow (y \wedge z) = (x \Rightarrow y) \wedge (x \Rightarrow z)$$

are satisfied.

Proof Suppose the equations hold. First note that it follows immediately from (iv) that the map $x \Rightarrow (-)$ is order preserving. Then if $z \leq x \Rightarrow y$ we have

$$z \wedge x \leq x \wedge (x \Rightarrow y) = x \wedge y \leq y.$$

Conversely, if $z \wedge x \leq y$, then

$$
\begin{aligned}
z &= z \wedge (x \Rightarrow z) && \text{by (iii)} \\
&\leq (x \Rightarrow x) \wedge (x \Rightarrow z) && \text{by (i)} \\
&= x \Rightarrow (x \wedge z) && \text{by (iv)} \\
&\leq x \Rightarrow y && \text{since } x \Rightarrow (-) \text{ is order preserving.}
\end{aligned}
$$

Conversely, suppose that L is a Heyting algebra. Since $x \wedge z \leq z$ always holds, it is clear that $x \Rightarrow x = 1$. And since $y \wedge x \leq y$, we have $y \leq x \Rightarrow y$, that is, $y \wedge (x \Rightarrow y) = y$. Now $x \wedge (x \Rightarrow y) \leq y$ by definition, and $x \wedge (x \Rightarrow y) \leq x$, so $x \wedge (x \Rightarrow y) \leq x \wedge y$. But $(x \wedge y) \wedge x \leq y$, whence $x \wedge y \leq x \Rightarrow y$, and $x \wedge y \leq x$, so $x \wedge y \leq x \wedge (x \Rightarrow y)$. Hence $x \wedge (x \Rightarrow y) = x \wedge y$. Finally, it is clear that $x \Rightarrow (-)$ is order-preserving, so that $x \Rightarrow (y \wedge z) \leq (x \Rightarrow y) \wedge (x \Rightarrow z)$. But

$$(x \Rightarrow y) \wedge (x \Rightarrow z) \wedge x = [x \wedge (x \Rightarrow y)] \wedge [x \wedge (x \Rightarrow z)]$$
$$\leq y \wedge z.$$

Hence $(x \Rightarrow y) \wedge (x \Rightarrow z) \leq x \Rightarrow (y \wedge z)$. $\qquad\qquad \square$

Any linearly ordered set with top and bottom elements is a Heyting algebra in which $x \Rightarrow y = \begin{cases} 1 & \text{if } x \le y \\ y & \text{if } y < x. \end{cases}$

A basic fact about *complete* Heyting algebras is that the following identity holds in them:

$$(*) \qquad x \wedge \bigvee_{i \in I} y_i = \bigvee_{i \in I} x \wedge y_i.$$

And conversely, in any complete lattice satisfying (*), defining the operation $\Rightarrow$ by $x \Rightarrow y = \bigvee \{z : z \wedge x \le y\}$ turns it into a Heyting algebra.

To prove this, we observe that in any complete Heyting algebra,

$$x \wedge \bigvee_{i \in I} y_i \le z \leftrightarrow \bigvee_{i \in I} y_i \le x \Rightarrow z$$

$$\leftrightarrow y_i \le x \Rightarrow z, \text{ all } i$$

$$\leftrightarrow y_i \wedge x \le z, \text{ all } i$$

$$\leftrightarrow \bigvee_{i \in I} x \wedge y_i \le z.$$

Conversely, if (*) is satisfied and $x \Rightarrow y$ is defined as above, then

$$(x \Rightarrow y) \wedge x \le \bigvee \{z : z \wedge x \le y\} \wedge x = \bigvee \{z \wedge x : z \wedge x \le y\} \le y.$$

So $z \le x \Rightarrow y \rightarrow z \wedge x \le (x \Rightarrow y) \wedge x \le y$. The reverse inequality is an immediate consequence of the definition.

In view of this result a complete Heyting algebra may also be defined to be a complete lattice satisfying (*). Complete Heyting algebras are also known as *frames*.

If X is a topological space, then the complete lattice $O(X)$ of open sets in X is a Heyting algebra. In $O(X)$ meet and join are just set-theoretic intersection and union, while the implication and pseudocomplementation operations are given by

$$U \Rightarrow V = \overset{\circ}{\overbrace{(X - U) \cup V}} \text{ and } U^* = \overset{\circ}{\overbrace{X - U}}.$$

A *subalgebra* of a Heyting algebra H is a sublattice of H, which is closed under Hs implication operation. A subalgebra of $O(X)$ for some topological space X is called an *algebra of opens*.

Let L be a bounded lattice. A *complement* for an element $a \in L$ is an element $b \in L$ satisfying $a \vee b = 1$ and $a \wedge b = 0$. In general, an element of a lattice may have more than one complement, or none at all. However, in a *distributive* lattice an element can have at most one complement. For if b, b' are complements of an element a of a distributive lattice, then $a \vee b = a \vee b' = 1$ and $a \wedge b = a \wedge b' = 0$. From this we deduce

$$b = b \vee 0 = b \vee (a \wedge b') = (b \vee a) \wedge (b \vee b') = 1 \vee (b \vee b') = b \vee b'.$$

Similarly $b' = b \vee b'$ so that $b = b'$.

In a Heyting algebra H the pseudocomplement a^* of an element a is not, in general, a complement for a. (Consider the Heyting algebra of open sets of a topological space.) But there is a simple necessary and sufficient condition on a Heyting algebra for all pseudocomplements to be complements: this is stated in the following.

Proposition 0.2 *The following conditions on a Heyting algebra H are equivalent:*

  (i) *pseudocomplements are complements, that is, $x \vee x^* = 1$ for all $x \in H$;*

  (ii) *pseudocomplementation is of order 2, that is, $x^{**} = x$ for all $x \in H$.*

Proof (i) $\rightarrow$ (ii). Assuming (i), we have

$$x^{**} = x^{**} \wedge 1 = x^{**} \wedge (x \vee x^*)$$
$$= (x^{**} \wedge x) \vee (x^{**} \wedge x^*)$$
$$= (x^{**} \wedge x) \vee 0 = (x^{**} \wedge x).$$

Therefore $x^{**} \leq x$ whence $x^{**} = x$.

(ii) $\rightarrow$ (i). We have $(x \vee x^*)^* = x^* \wedge x^{**} = 0$, so assuming (ii) gives $x \vee x^* = (x \vee x^*)^{**} = 0^* = 1$. $\square$

We now define a *Boolean algebra* to be a Heyting algebra satisfying either of the equivalent conditions, Proposition 0.2 (i) or (ii). The following identities accordingly hold in any Boolean algebra:

$$x \vee y = y \vee x, \quad x \wedge y = y \wedge x$$
$$x \vee (y \vee z) = (x \vee y) \vee z, \quad x \wedge (y \wedge z) = (x \wedge y) \wedge z$$
$$(x \vee y) \wedge y = y, \quad (x \wedge y) \vee y = y$$
$$x \wedge (y \vee z) = (x \wedge y) \vee (x \wedge z), \quad x \vee (y \wedge z) = (x \vee y) \wedge (x \vee z).$$
$$x \vee x^* = 1, \quad x \wedge x^* = 0.$$
$$(x \vee y)^* = x^* \wedge y^*, \quad (x \wedge y)^* = x^* \vee y^*$$
$$x^{**} = x.$$

It is easy to show that in any Boolean algebra $x \Rightarrow y = x^* \vee y$. In a *complete* Boolean algebra we have the following identities:

$$\left(\bigvee_{i \in I} x_i\right)^* = \bigwedge_{i \in I} x_i^*, \quad \left(\bigwedge_{i \in I} x_i\right)^* = \bigvee_{i \in I} x_i^*,$$

$$x \wedge \bigvee_{i \in I} y_i = \bigvee_{i \in I} (x \wedge y_i), \quad x \vee \bigwedge_{i \in I} y_i = \bigwedge_{i \in I} (x \vee y_i).$$

Calling a lattice *complemented* if it is bounded and each of its elements has a complement, we can characterize Boolean algebras alternatively as *complemented distributive lattices*. For we have already shown that every Boolean algebra is distributive and complemented. Conversely, given a complemented distributive lattice L, write a^c for the (unique) complement of an element a; it is then easily shown that defining implication by $x \Rightarrow y = x^c \vee y$ turns L into a Heyting algebra in which x^* coincides with x^c, so that L is Boolean.

The meet, join, and complementation operations in a Boolean algebra are called its *Boolean operations*. A *subalgebra* of a Boolean algebra B is a nonempty subset closed under Bs Boolean operations. Clearly a subalgebra of a Boolean algebra B is itself a Boolean algebra with the same top and bottom elements as those of B.

Examples of Boolean algebras (i) The linearly ordered set $2 = \{0, 1\}$ with $0 < 1$ is a complete Boolean algebra, the *two-element algebra*.

(ii) The power set lattice PA of any set A is a complete Boolean algebra. A subalgebra of a power set algebra is called a *field of sets*.

(iii) Let $F(A)$ consist of all finite subsets and all complements of finite subsets of a set A. With the partial ordering of inclusion, $F(A)$ is a field of sets called the *finite–cofinite algebra* of A.

(iv) Let X be a topological space, and let $C(X)$ be the family of all simultaneously closed and open (clopen) subsets of X. With the partial ordering of inclusion, $C(X)$ is a Boolean algebra called the *clopen algebra* of X.

(v) A subset U of a topological space X is said to be *regular open*[2] if $\overset{\circ}{\overline{U}} = U$. The family $\mathrm{RO}(X)$ of all regular open subsets of X, partially ordered by inclusion, is a complete Boolean algebra[3]—the *regular open algebra* of X—in which 0 is $\varnothing$, 1 is X, and for $U, V \in \mathrm{RO}(X), U \vee V = \overset{\circ}{\overline{U \cup V}}, U \wedge V = U \cap V$

---

[2] In the Euclidean plane, regular open sets are those lacking 'cracks' or 'pinholes'.

[3] For a proof of this, see, for example, Halmos 1963, section 7, Lemma 1.

and $U^* = X - \overline{U}$. Moreover, for $\{Ui : i \in I\} \subseteq \mathrm{RO}(X)$,

$$\bigvee_{i \in I} U_i = \overline{\bigcup_{i \in I} U_i}^{\,\circ} \qquad \bigwedge_{i \in I} U_i = \overset{\circ}{\overparen{\bigcap_{i \in I} U_i}}.$$

An element a of a Heyting algebra H is said to be *regular* if $a = a^{**}$. A regular open set in a topological space X is then precisely a regular element of $O(X)$. Clearly a Heyting algebra is a Boolean algebra if and only if each of its elements is regular, so that $O(X)$ is a Boolean algebra if and only if each of its open subsets is regular open. In particular, if $\mathbb{R}$ is the space of real numbers with its usual topology, $O(\mathbb{R})$ is not a Boolean algebra.

Let B be the set of regular elements of H; it can be shown that B, with the partial ordering inherited from H, is a Boolean algebra—the *regularization* of H—in which the operations $\wedge$ and $*$ coincide with those of H, but[4] $\vee_B = (\vee_H)^{**}$. If H is complete, so is B; the operation $\bigwedge$ in B coincides with that in H while $\bigvee_B = (\bigvee_H)^{**}$.

It follows that the regular open algebra $\mathrm{RO}(X)$ coincides with the regularization of $O(X)$.

Filters, ideals, and homomorphisms

Let L be a (bounded) distributive lattice. A *filter (ideal)* in L is a subset F such that $1 \in F; 0 \notin F; x, y \in F \rightarrow x \wedge y \in F; x \in F$ and $x \leq y \rightarrow y \in F$ (resp. $0 \in I; 1 \notin I; x, y \in I \rightarrow x \vee y \in I; x \in I$ and $y \leq x \rightarrow y \in I$.) Clearly a lattice is trivial iff it contains no filters or ideals. A subset X of L has the *finite meet property* (f.m.p) if the meet of any nonempty finite subset of X is $\neq 0$. Clearly any subset of a filter has the f.m.p.; conversely, any subset X having the f.m.p. is included in a filter, namely

$$X^+ = \{y \in L : \exists x_1 \in X \ldots \exists x_n \in X[x_1 \wedge \cdots \wedge x_n \leq y]\}.$$

X^+ is the least filter containing X—the filter *generated* by X. In particular for $a \neq 0$ the filter $\{a\}^+ = \{x : a \leq x\}$ is the least filter containing a; it is called the *principal* filter generated by a.

A filter F in L is *prime* if it satisfies the condition $x \vee y \in F \rightarrow x \in F$ or $y \in F$: if L is a Boolean algebra, this is easily shown to be equivalent to the condition $\forall x[x \in F$ or $x^* \in F]$. A filter maximal under inclusion is called an *ultrafilter*. It is readily shown that a filter F is an ultrafilter iff it satisfies the condition $\forall x[\forall y \in F(x \wedge y \neq 0 \rightarrow x \in F]$; in a Heyting algebra this condition is equivalent to $\forall x[x \in F \vee x^* \in F]$. It follows that, in a Boolean algebra, prime

---

[4]Here we write $\vee_B$, $\vee_H$ for the join operations in B and H respectively, with similar conventions employed below.

filters and ultrafilters coincide. Every ultrafilter in a distributive lattice is prime. More generally, in a distributive lattice we have the

Proposition 0.3 *Let F be a filter in a distributive lattice L, and b an element of L such that $b \notin F$. Then there is a filter containing F maximal under inclusion with respect to the property of not containing b. Any such filter is prime.*

Proof The family $\mathscr{F}$ of filters in L containing F but not b is readily shown to be closed under unions of chains, and so by Zorn's lemma has a maximal element. Let M be a maximal element of $\mathscr{F}$ and suppose that $c \vee d \in M$. Consider the sets

$$M_c = \{y \in L : \exists x \in M \ \ y \geq x \wedge c\}, \quad M_d = \{y \in L : \exists x \in M \ \ y \geq x \wedge d\}.$$

If $b \in M_c \cap M_d$, then there are $x, y \in M$ such that $b \geq x \wedge c$ and $b \geq y \wedge d$. It follows that

$$b \geq (x \wedge c) \vee (y \wedge d) = (x \vee y) \wedge (x \vee d) \wedge (c \vee y) \wedge (c \vee d) \in M,$$

whence $b \in M$ contrary to assumption. Therefore $b \notin M_c$ or $b \notin M_d$. In the first case F_c is then a filter $\supseteq M$ and so $M = M_c$ by the maximality of M, whence $c \in m$. Similarly, in the second case, $d \in M$. Accordingly M is prime. $\qquad\square$

This proposition has the following immediate consequences

Corollary 0.4 *For any elements a, b of a distributive lattice such that $a \not\leq b$ there is a prime filter containing a but not b.*

Corollary 0.5 *Each filter in a distributive lattice L with top and bottom elements is included in an ultrafilter.*

If L and L' are distributive lattices, a map $h : L \to L'$ is a *lattice homomorphism* if it satisfies: $h(0) = 0, h(1) = 1, h(x \wedge y) = h(x) \wedge h(y), h(x \vee y) = h(x) \vee h(y)$ for arbitrary $x, y \in L$. Lattice homomorphisms are clearly order preserving. A homomorphism $h : L \to L'$ is *complete* if, for any $X \subseteq L$ such that $\vee X$ exists in L, $\vee h[x]$ exists in L' and equals $h(\vee X)$. If $h : L \to L'$ is a homomorphism, and L' nontrivial, the set $h^{-1}[1] = \{x : h(x) = 1\}$—the *hull* of f—is a filter, and $h^{-1}[0] = \{x : h(x) = 0\}$—the *kernel* of f—an ideal in L. When $h : L \to 2$, the filter $h^{-1}[1]$ is a prime filter. Conversely, each prime filter P in L determines a homomorphism $h : L \to 2$ defined by $h(x) = 1$ iff $x \in P$.

If H and H' are Heyting algebras, a lattice homomorphism $h : H \to H'$ is an *algebra homomorphism* if $h(x \Rightarrow y) = h(x) \Rightarrow h(y)$ for arbitrary $x, y \in H$. When $h : H \to 2$, the filter $h^{-1}[1]$ is an ultrafilter. Conversely, each ultrafilter U in H determines an algebra homomorphism $h : H \to 2$ defined by $h(x) = 1$ iff $x \in U$.

It is easily verified that, for Boolean algebras B, B', a map $h : B \to B'$ is a(n) (algebra) homomorphism iff it satisfies either of the two equivalent conditions: (i) $h(x \wedge y) = h(x) \wedge h(y), h(x^*) = h(x)^*$ for all $x, y \in B$; (ii) $h(x \vee y) = h(x) \vee h(y), h(x^*) = h(x)^*$ for all $x, y \in B$. Moreover, h is injective iff $h^{-1}[1] = 1$, or, equivalently, if $h^{-1}[0] = 0$.

If I is an ideal in a Boolean algebra B, we define the *quotient algebra* B/I as follows. Introduce the relation $\sim_I$ on B by $a \sim_I b$ iff, for some $x \in I, a \vee x = b \vee x$. It is easily verified that $\sim_I$ is a congruence relation on B, that is, an equivalence relation satisfying the condition:

$$\text{If } a \sim_I a', b \sim_I b', \text{ then } a \vee b \sim_I a' \vee b', \; a \wedge b \sim_I a' \wedge b', \; a^* \sim_I a'^*.$$

Writing a/I for the $\sim_I$-equivalence class of a, it follows that the set $B/I = \{a/I : a \in B\}$ can be assigned the structure of a Boolean algebra by defining $a/I \wedge b/I = (a \wedge b)/I, a/I \vee b/I = (a \vee b)/I, (a/I)^* = a^*/I$. The top and bottom elements of B/I are $1/I$ and $0/I$. The map $h : B \to B/I$ given by $h(a) = a/I$ is then a homomorphism onto B/I called the *canonical homomorphism*. Clearly the kernel of h is I.

Similarly, starting with a filter F in B and defining the congruence relation $\approx_F$ on B by $a \approx_F b$ iff, for some $x \in F, a \wedge x = b \wedge x$ yields the quotient algebra B/F: in this case the hull of the corresponding canonical homomorphism $B \to B/F$ is F.

A bijective homomorphism is called an *isomorphism*; two lattices, Heyting, or Boolean algebras are *isomorphic* if there exists an isomorphism between them.

If B and B' are Boolean algebras, a homomorphism $h : B \to B'$ is *complete* if, for any $X \subseteq B$ such that $\bigvee X$ exists in B, $\bigvee\{h(x) : x \in X\}$ exists in B' and equals $h(\bigvee x)$. An isomorphism of a Boolean algebra B with itself is called an *automorphism;* clearly any automorphism is a complete homomorphism. The set of all automorphisms of B forms a group under function composition: this group is called the *group of automorphisms* of B.

In the sequel we shall require a result concerning the existence of special sorts of ultrafilters in Boolean algebras, the *Rasiowa–Sikorski Theorem*. Let S be a family of subsets of a Boolean algebra B, each member X of which has a join $\bigvee X$. An ultrafilter U in B is said to be *S-complete* if for all $X \in S$ we have $\bigvee X \in U \to X \cap U \neq \varnothing$. Equivalently, U is S-complete if whenever $X \subseteq U$ and $\{x^* : x \in X\} \in S$, then $\bigwedge X \in U$.

Rasiowa–Sikorski Theorem 0.6 *If each member of a countable family S of subsets of a Boolean algebra B has a join, then for each $a \neq 0$ in B there is an S-complete ultrafilter in B containing a.*

Proof Enumerate S as $\{T_n : n \in \omega\}$ and write $t_n = \bigvee T_n$. We define by recursion a sequence $b_0, b_1, \ldots$ of elements of B such that, for each $n \in \omega, b_n \in T_n$ and

$$a \wedge (t_0^* \vee b_0) \wedge \cdots \wedge (t_n^* \vee b_n) \neq 0.$$

Given $n \in \omega$, suppose that for each $m < n$, b_m has been found so as to satisfy the specified conditions. If $n = 0$, let $c = a$, and if $n > 0$ let

$$c = a \wedge (t_0^* \vee b_0) \wedge \cdots \wedge (t_{n-1}^* \vee b_{n-1}).$$

Then if $n = 0$ we have $c \neq 0$ by assumption, and if $n > 0$ we have $c \neq 0$ by inductive hypothesis. Suppose now, for contradiction's sake, that $c \wedge (t_n^* \vee b) = 0$ for all $b \in T_n$. Then $0 = (c \wedge t_n^*) \vee (c \wedge b)$, so that $c \wedge t_n^* = 0$ and $c \wedge b = 0$, whence $c \leq b^*$ for all $b \in T_n$. Hence

$$c \leq \bigwedge \{b^* : b \in T_n\} = \left(\bigvee T_n\right)^* = t_n^*.$$

So $c = c \wedge t_n^* = 0$ contradicting the inductive hypothesis.

Accordingly b_n can be found to satisfy the specified conditions, and so such a b_n can be found for each $n \in \omega$. Then the set $\{a\} \cup \{t_n^* \vee b_n : n \in \omega\}$ has the f.m.p. and is therefore included in an ultrafilter U in B. U certainly contains a; we show that U is S-complete. If $\bigvee T_n = t_n \in U$, then, since $t_n^* \vee b_n \in U$ by construction, it follows that $b_n = t_n \wedge b_n = t_n \wedge (t_n^* \vee b_n) \in U$, so that $T_n \cap U \neq \varnothing$. $\qquad \square$

Representation theorems for distributive lattices

We state and prove three representation theorems:

Theorem 0.7 *Any distributive lattice is isomorphic to a lattice of sets.*

Theorem 0.8 *Any Heyting algebra is isomorphic to an algebra of opens.*

Theorem 0.9 (The Stone Representation Theorem) *Any Boolean algebra is isomorphic to a field of sets.*

The proofs of these theorems all rest on the introduction of the set $\mathfrak{P}(L)$ of prime filters in a distributive lattice L. There is a canonical map $p : L \to P(\mathfrak{P}(L))$ defined by $p(x) = \{P \in \mathfrak{P}(L) : x \in P\}$ for $x \in L$. It is easy to verify that p is a lattice homomorphism; it follows from corollary 0.4 that p is injective, and hence an isomorphism of L with the lattice of sets $\{p(x) : x \in L\}$. This proves Theorem 0.7.

To prove Theorem 0.8 we start with a Heyting algebra H and impose a certain topology on $\mathfrak{P}(H)$. As a base for this topology we take the family $\{p(x) : x \in H\}$. Write X for the resulting topological space. We show that, for $x, y \in H$,

$$(*) \qquad p(x \Rightarrow y) = \overset{\circ}{\overline{[X - p(x)] \cup p(y)}} = p(x) \Rightarrow p(y).$$

To prove (*) we first observe that

$$(\text{**}) \quad \overset{\circ}{\overbrace{P \in [X - p(x)] \cup p(y)}} \leftrightarrow \exists z[P \in p(z) \subseteq [X - p(x)] \cup p(y)]$$

$$\leftrightarrow \exists z \in P[\forall Q \in X[z \in Q \to [Q \notin p(x) \text{ or } Q \in p(y)]]]$$

$$\leftrightarrow \exists z \in P[\forall Q \in X[z \in Q \to [x \in Q \to y \in Q]]]$$

$$\leftrightarrow \exists z \in P[\forall Q \in X[z \in Q \text{ and } x \in Q \to y \in Q]].$$

Next, we note that

$$(\text{***}) \quad z \leq x \Rightarrow y \leftrightarrow z \wedge x \leq y \leftrightarrow \quad \forall Q \in X[z \in Q \text{ and } x \in Q \to y \in Q].$$

For if $z \wedge x \leq y$, and $Q \in X$ satisfies $z \in Q$ and $x \in Q$, then $z \wedge x \in Q$, whence $y \in Q$. Conversely, suppose that, for every $Q \in X, z \in Q$ and $x \in Q \to y \in Q$. If $z \wedge x \not\leq y$, then by Corollary 0.4 there is $Q \in X$ containing $z \wedge x$ and hence both z and x but not y, contradicting assumption. Hence $z \wedge x \leq y$.

It now follows from (**) and (***) that

$$\overset{\circ}{\overbrace{P \in [X - p(x)] \cup p(y)}} \leftrightarrow \exists z \in P[z \leq x \Rightarrow y] \leftrightarrow x \Rightarrow y \in P \leftrightarrow P \in p(x \Rightarrow y),$$

which immediately yields (*).

Therefore p is an algebra homomorphism of H into $O(X)$. We have already seen that it is injective, so it is an isomorphism of H with the algebra of opens $\{p(x) : x \in H\}$. This proves Theorem 0.8.

Finally we prove Theorem 0.9. Let B be a Boolean algebra. We have already shown that the canonical homomorphism $p : B \to P(\mathfrak{P}(B))$ is an isomorphism of B with the lattice of sets $\tilde{B} = \{p(x) : x \in L\}$. It is readily seen that, since B is a Boolean algebra, p preserves complements, so that $\tilde{B}$ is a field of sets and p an isomorphism between the Boolean algebras B and $\tilde{B}$.

Connections with logic

Heyting and Boolean algebras have close connections with *intuitionistic* and *classical* logic,[5] respectively.

---

[5] For accounts of both systems of logic, see, for example, Bell and Machover 1977.

Intuitionistic first-order logic has the following axioms and rules of inference.
Axioms

$$\varphi \to (\psi \to \varphi)$$
$$[\varphi \to (\psi \to \chi) \to [(\varphi \to \psi) \to (\varphi \to \chi)]$$
$$\varphi \to (\psi \to \varphi \wedge \psi)$$
$$\varphi \wedge \psi \to \varphi \qquad \varphi \wedge \psi \to \psi$$
$$\varphi \to \varphi \vee \psi \qquad \psi \to \varphi \vee \psi$$
$$[\varphi \to (\psi \to \chi) \to [(\varphi \to \psi) \to (\varphi \to \chi)]$$
$$(\varphi \to \chi) \to [(\psi \to \chi) \to (\varphi \vee \psi \to \chi)]$$
$$(\varphi \to \psi) \to [(\varphi \to \neg\psi) \to \neg\varphi]$$
$$\neg\varphi \to (\varphi \to \psi)$$

$$\varphi(t) \to \exists x \varphi(x) \qquad \forall x \varphi(x) \to \varphi(y) \ (x \text{ free in } \varphi \text{ and } t \text{ free for } x \text{ in } \varphi)$$
$$x = x \qquad \varphi(x) \wedge x = y \to \varphi(y).$$

Rules of Inference

$$\frac{\varphi, \varphi \to \psi}{\psi}$$

$$\frac{\psi \to \varphi(x)}{\psi \to \forall x \varphi(x)} \qquad \frac{\varphi(x) \to \psi}{\exists x \varphi(x) \to \psi}.$$

$$(x \text{ not free in } \psi)$$

Classical first-order logic is obtained by adding to the intuitionistic system the
rule of inference

$$\frac{\neg\neg\varphi}{\varphi}.$$

In intuitionistic logic none of the classically valid logical schemes

LEM (law of excluded middle) $\varphi \vee \neg\varphi$
LDN (law of double negation) $\neg\neg\varphi \to \varphi$
DEM (de Morgan's law) $\neg(\varphi \wedge \psi) \to \neg\varphi \vee \neg\psi$

are derivable. However LEM and LDN are intuitionistically equivalent and DEM
is intuitionistically equivalent to the weakened law of excluded middle:

WLEM $\neg\varphi \vee \neg\neg\varphi.$

Also the weakened form of LDN for negated statements,

WLDN $\neg\neg\neg\varphi \to \neg\varphi$

is intuitionistically derivable. It follows that any formula intuitionistically equivalent to a negated formula satisfies the LDN.

Heyting algebras are associated with theories in intuitionistic logic in the following way. Given a consistent theory T in an intuitionistic propositional or first-order language **L**, define the equivalence relation $\approx$ on the set of formulas of **L** by $\varphi \approx \psi$ if $T \vdash \varphi \leftrightarrow \psi$. For each formula φ write $[\varphi]$ for its $\approx$-equivalence class. Now define the relation $\leq$ on the set $H(T)$ of $\approx$-equivalence classes by $[\varphi] \leq [\psi]$ if and only if $T \vdash \varphi \to \psi$. Then $\leq$ is a partial ordering of $H(T)$ and the partially ordered set $(H(T), \leq)$ is a Heyting algebra in which $[\varphi] \Rightarrow [\psi] = [\varphi \to \psi]$, with analogous equalities defining the meet and join operations, 0, and 1. $H(T)$ is called the the Heyting algebra *determined by* T. It can be shown that Heyting algebras of the form $H(T)$ are typical in the sense that, for any Heyting algebra L, there is a propositional intuitionistic theory T such that L is isomorphic to $H(T)$. Accordingly Heyting algebras may be identified as the algebras of intuitionistic logic.

Similarly, starting with a consistent theory T in a classical propositional or first-order language, the associated algebra $B(T)$ is a Boolean algebra known as the *Lindenbaum algebra* of T. Again, it can be shown that any Boolean algebra is isomorphic to $B(T)$ for a suitable classical theory T.

As regards semantics, Heyting algebras and Boolean algebras have corresponding relationships with intuitionistic, and classical, propositional logic, respectively. Thus, suppose given a propositional language $\mathscr{L}$; let $\mathfrak{P}$ be its set of propositional variables. Given a map $f\colon \mathfrak{P} \to H$ to a Heyting algebra H, we extend f to a map $\varphi \mapsto [\![\varphi]\!]$ of the set of formulas of $\mathscr{L}$ to H by:

$$[\![\varphi \wedge \psi]\!] = [\![\varphi]\!] \wedge [\![\psi]\!] \quad [\![\varphi \vee \psi]\!] = [\![\varphi]\!] \vee [\![\psi]\!]$$
$$[\![\varphi \Rightarrow \psi]\!] = [\![\varphi]\!] \Rightarrow [\![\psi]\!] \quad [\![\neg\varphi]\!] = [\![\varphi]\!]^{*}.$$

A formula φ is said to be *Heyting valid*—written $\vdash \varphi$—if $[\![\varphi]\!] = 1$ for any such map f. It can then be shown that φ is Heyting valid iff $\vdash \varphi$ in the intuitionistic propositional calculus, that is, iff φ is provable from the propositional axioms listed above.

Similarly, if we define the notion of Boolean validity by restricting the definition of Heyting validity to maps into Boolean algebras, then it can be shown that a formula is Boolean valid iff it is provable in the *classical* propositional calculus.

Finally, again as regards semantics, *complete* Heyting and Boolean algebras are related to intuitionistic, and classical *first-order* logic, respectively. To be precise, let $\mathscr{L}$ be a first-order language whose sole extralogical symbol is a binary predicate symbol P. A *Heyting algebra–valued $\mathscr{L}$-structure* is a quadruple $\mathbf{M} = (M, eq, Q, H)$, where M is a nonempty class, H is a complete Heyting algebra and eq and Q are maps $M \times M \to H$ satisfying, for all $m, n, m', n' \in M$,

$$eq(m, m) = 1, eq(m, n) = eq(n, m), eq(m, n) \wedge eq(n, n') \leq eq(m, n'),$$

$$Q(m, n) \wedge eq(m, m') \leq Q(m', n), Q(m, n) \wedge eq(n, n') \leq Q(m, n').$$

For any formula φ of $\mathscr{L}$ and any finite sequence $\boldsymbol{x} = \langle x_1, \ldots, x_n \rangle$ of variables of $\mathscr{L}$ containing all the free variables of φ, we define for any Heyting-valued $\mathscr{L}$-structure $\mathbf{M}$ a map

$$[\![\varphi]\!]^{\mathbf{M}\boldsymbol{x}} : M^n \to H$$

recursively as follows:

$$[\![x_p = x_q]\!]^{\mathbf{M}\boldsymbol{x}} = \langle m_1, \ldots, m_n \rangle \mapsto eq(m_p, m_q),$$

$$[\![Px_p x_q]\!]^{\mathbf{M}\boldsymbol{x}} = \langle m_1, \ldots, m_n \rangle \mapsto Q(m_p, m_q),$$

$$[\![\varphi \wedge \psi]\!]^{\mathbf{M}\boldsymbol{x}} = [\![\varphi]\!]^{\mathbf{M}\boldsymbol{x}} \wedge [\![\psi]\!]^{\mathbf{M}\boldsymbol{x}}, \text{ and similar clauses for the other connectives,}$$

$$[\![\exists y \varphi]\!]^{\mathbf{M}\boldsymbol{x}} = \langle m_1, \ldots, m_n \rangle \mapsto \bigvee_{m \in M} [\![\varphi(y/u)]\!]^{\mathbf{M}u\boldsymbol{x}}(m, m_1, \ldots, m_n)$$

$$[\![\forall y \varphi]\!]^{\mathbf{M}\boldsymbol{x}} = \langle m_1, \ldots, m_n \rangle \mapsto \bigwedge_{m \in M} [\![\varphi(y/u)]\!]^{\mathbf{M}u\boldsymbol{x}}(m, m_1, \ldots, m_n)$$

Call φ $\mathbf{M}$-*valid* if $[\![\varphi]\!]^{\mathbf{M}\boldsymbol{x}}$ is identically 1, where $\boldsymbol{x}$ is the sequence of all free variables of φ. Then it can be shown that φ is $\mathbf{M}$-*valid for all $\mathbf{M}$ iff φ is provable in intuitionistic first-order logic*. This is the *algebraic completeness theorem* for intuitionistic first-order logic.

Similarly, if we carry out the same procedure, replacing complete Heyting algebras with complete Boolean algebras, one can prove the corresponding algebraic completeness theorem for classical first-order logic, namely, a first-order formula is valid in every Boolean-valued structure iff it is provable in classical first-order logic.

$$1$$

BOOLEAN-VALUED MODELS OF SET THEORY:
FIRST STEPS

Basic set theory

Although it assumed that readers of this book will be familiar with the
development of axiomatic set theory up to the consistency of the axiom of choice
and the generalized continuum hypothesis (e.g., see Cohen 1966; Drake 1974; Bell
and Machover 1977; Kunen 1980), we begin with a brief review of the notions
from set theory that we shall need.

The *language of set theory* is a first-order language $\mathcal{L}$ with equality, which also
includes a binary predicate symbol $\in$ (membership). The *individual* variables
$v_0, v_1, \ldots, x, y, z, \ldots$ of $\mathcal{L}$ are understood to range over *sets*, but we shall also
permit the formation of *class terms* $\{x\colon \varphi(x)\}$ for each formula $\varphi(x)$. The term
$\{x\colon \varphi(x)\}$ is understood to denote 'the class of all (sets) x such that $\varphi(x)$'; a
term of this form will be called simply a (*definable*) *class*. We assume that
classes satisfy *Church's scheme:*

$$\forall y[y \in \{x\colon \varphi(x)\} \leftrightarrow \varphi(y)].$$

We shall employ the standard set-theoretic abbreviations, in particular the
following:

$\exists! x \varphi(x)$	'there is a unique x such that $\varphi(x)$'
$\mathrm{dom}(f)$	domain of f
$\mathrm{ran}(f)$	range of f
Px or $P(x)$	power set of x
$F \mid X$	restriction of F to X
$F[X]$	image of X under F
$\langle x, y \rangle$	ordered pair of x, y
x^y	set of all maps of y into x
$\mathrm{Fun}(f)$	'f is a function'
$\mathrm{Ord}(x)$	'x is an ordinal'
$\mathrm{L}(x)$	'x is constructible'
$\varnothing = \{x\colon x \neq x\}$	the *empty set*
$V = \{x\colon x = x\}$	the *universe of sets*
$\mathrm{ORD} = \{x\colon \mathrm{Ord}(x)\}$	the *class of ordinals*
$L = \{x\colon \mathrm{L}(x)\}$	the *constructible universe*.

We shall use lower case Greek letters $\alpha, \beta, \gamma, \xi, \eta$ as variables ranging over ordinals.

For our purposes *Zermelo–Fraenkel set theory* (ZF) is the theory in $\mathcal{L}$ based on the following axioms (1)–(7):

(1) *Extensionality*

$$\forall x \forall y [\forall z (z \in x \leftrightarrow z \in y) \to x = y].$$

(2) *Separation*

$$\forall u \exists v \forall x [x \in v \leftrightarrow x \in u \wedge \varphi(x)],$$

where v is not free in the formula $\varphi(x)$.

(3) *Replacement*

$$\forall u [\forall x \in u \, \exists y \, \varphi(x, y) \to \exists v \, \forall x \in u \, \exists y \in v \, \varphi(x, y)],$$

where v is not free in the formula $\varphi(x, y)$.

(4) *Union*

$$\forall u \, \exists v \, \forall x [x \in v \leftrightarrow \exists y \in u (x \in y)].$$

(5) *Power set*

$$\forall u \, \exists v \, \forall x [x \in v \leftrightarrow \forall y \in x (y \in u)].$$

(6) *Infinity*

$$\exists u [\varnothing \in u \wedge \forall x \in u \, \exists y \in u (x \in y)].$$

(7) *Regularity*

$$\forall x [\forall y \in x \varphi(y) \to \varphi(x)] \to \forall x \varphi(x)],$$

where y is not free in the formula $\varphi(x)$.

Remarks 1 The forms of the axioms of replacement (3) and regularity (7), which have been chosen so as to allow easy verification in the models we shall construct, differ from their customary forms, viz.

$$(3^*) \qquad \forall u [\forall x \in u \, \exists! y \, \varphi(x, y) \to \exists v \, \forall y [y \in v \leftrightarrow \exists x \in u \, \varphi(x, y)]].$$

and

$$(7^*) \qquad \forall u [u \neq \varnothing \to \exists x \in u (x \cap u = \varnothing)].$$

It is not difficult to show that the replacing of (3) by (3*) and (7) by (7*) yields a system equivalent in strength.

2 The usual axiom of pairing has been omitted from our presentation of ZF because it is derivable from the axioms of replacement, power set, and separation.

The *axiom of choice* (AC) is the sentence

$$\forall u \exists f [\mathrm{Fun}(f) \wedge \mathrm{dom}(f) = u \wedge \forall x \in u [x \neq \varnothing \to f(x) \in x]].$$

Over ZF the axiom of choice is equivalent to *Zorn's lemma*, namely, the assertion that, for any partially ordered set P in which each chain has a supremum (or merely an upper bound), P has a maximal element.

The theory ZF + AC is written ZFC.

We conceive of *ordinals* in such a way that each ordinal coincides with the set of its predecessors. A *cardinal* is an ordinal not equipollent (bijective) with any smaller ordinal. Customarily we use the Greek letters κ, λ to denote *infinite* cardinals. We assume that the infinite cardinals are enumerated in a sequence $\aleph_0, \aleph_1, \ldots, \aleph_\alpha, \ldots$ or, when considered as ordinals, $\omega_0, \omega_1, \ldots, \omega_\alpha, \ldots$. The next cardinal after κ is written κ^+; thus, if $\kappa = \aleph_\alpha$, then $\kappa^+ = \aleph_{\alpha+1}$. The cardinality of a set x—that is, the least cardinal equipollent with x—is denoted by $|x|$. When κ and λ are cardinals we usually write κ^λ for $|\kappa^\lambda|$. (The context will make it clear whether a given occurrence of κ^λ is intended to mean the set of maps from λ to κ or its cardinality.) Recall that $|Px| = 2^{|x|}$ for any set x. The *continuum hypothesis* (CH) is the assertion $2^{\aleph_0} = \aleph_1$; the *generalized continuum hypothesis* (GCH) is the assertion that, for all infinite cardinals κ, $2^\kappa = \kappa^+$. An infinite cardinal κ is said to be *regular* if whenever $|I| < \kappa$ and $\{\lambda_i : i \in I\} \subseteq \kappa$ then $\sum_{i \in I} \lambda_i < \kappa$. Notice that κ^+ is always regular. Also, if GCH holds and κ is regular, then $\kappa^\lambda = \kappa$ for any cardinal $\lambda < \kappa$.

For any infinite cardinal κ there is a *canonical bijection* between κ and $\kappa \times \kappa$ defined as follows. We well-order $\kappa \times \kappa$ by placing $\langle \xi, \eta \rangle$ before $\langle \xi', \eta' \rangle$ provided the ordered triple $\langle \max(\xi, \eta), \xi, \eta \rangle$ lexicographically precedes $\langle \max(\xi', \eta'), \xi', \eta' \rangle$. Then (see theorem 5.1 of Drake 1974) the canonical bijection of κ with $\kappa \times \kappa$ is an order isomorphism. If for each $\xi < \kappa$ we write $\langle \beta_\xi, \gamma_\xi \rangle$ for the element of $\kappa \times \kappa$ paired with ξ under the canonical bijection, it is not hard to show that $\beta_\xi \leq \xi$.

The *axiom of constructibility* is the sentence $\forall x L(x)$, or equivalently $V = L$. We recall the celebrated results of Gödel that, if ZF is consistent, so is $ZF + V = L$, and that GCH and AC are both provable in the latter theory.

We shall need some facts about *induction* and *recursion* on *well-founded relations*. A class R of ordered pairs is said to be *well-founded* if for each set u the class $\{x : xRu\} = Ru$ is a set and each nonempty set u has an element x such that yRx for no $y \in u$. The second condition is equivalent (assuming AC) to the assertion that for no sequence of sets $x_0, x_1, \ldots$ do we have $x_{n+1} \in x_n$ for all n.

Given a well-founded relation R, the *principle of induction on R* is the assertion,

$$\forall x [\forall y (yRx \to \varphi(y)) \to \varphi(x)] \to \forall x \varphi(x)$$

for an arbitrary formula $\varphi(x)$. The *principle of recursion on R* is the assertion that if F is any class of ordered pairs, which defines a single-valued mapping of V into V (such a class is called a *function* on V), then there is a function G on V such that

$$\forall u [G(u) = F(\langle u, G|Ru \rangle)].$$

The principles of induction and recursion on well-founded relations are provable in ZF as schemes.

It follows from the axiom of regularity that the membership relation is well-founded, so we may define the sets V_α by recursion as follows:

$$V_\alpha = \{x \colon \exists \xi < \alpha [x \subseteq V_\xi]\}.$$

The axiom of regularity implies that $\forall x \exists \alpha (x \in V_\alpha)$, so we may define a function $\mathrm{rank}(x)$ by setting

$$\mathrm{rank}(x) = \ \text{least } \alpha \text{ such that } x \in V_{\alpha+1}.$$

The relation $\mathrm{rank}(x) < \mathrm{rank}(y)$ is clearly well-founded, so we may deduce the *principle of induction on rank*:

$$\forall x [\forall y (\mathrm{rank}(y) < \mathrm{rank}(x) \to \varphi(y)) \to \varphi(x)] \to \forall x \varphi(x).$$

Finally, a few remarks on *models* of set theory. A(n) ($\mathcal{L}$-) *structure* is a pair $\langle M, E \rangle$ where M is a nonempty set and $E \subseteq M \times M$. *We shall customarily identify a given structure with its underlying set M and write 'M' for both.* If E is a well-founded relation, the structure M is said to be *well-founded*. A *transitive $\in$-structure* is a structure of the form $\langle M, \in | M \rangle$ in which M is a *transitive* set (i.e. satisfies $x \in y \in M \to x \in M$) and $\in | M = \{\langle x, y \rangle \in M \times M \colon x \in y\}$ is the $\in$-relation restricted to M. A *transitive $\in$-model* of ZF or ZFC is a transitive $\in$-structure, which is a model of ZF or ZFC. *Mostowski's collapsing lemma* is the assertion, provable in ZF, that if $\langle M, E \rangle$ is a well-founded structure which is a model of the axiom of extensionality, then there is a unique isomorphism h of $\langle M, E \rangle$ onto a transitive $\in$-structure, where h satisfies $h(x) = \{h(y) : yEx\}$ for $x \in M$. This transitive $\in$-structure is called the *transitive collapse* of $\langle M, E \rangle$.

If $\varphi(v_1, \ldots, v_n)$ is an $\mathcal{L}$-formula, M an $\mathcal{L}$-structure, and $a_1, \ldots, a_n \in M$, we write $M \models \varphi[a_1, \ldots, a_n]$ for '$a_1, \ldots, a_n$ satisfies φ in M'. If M is a model of ZFC and t a closed term of $\mathcal{L}$, we write $t^{(M)}$ for the *interpretation* of t in M. This is defined as follows: if t is a *set*, then $t^{(M)}$ is the unique $a \in M$ for which $M \models (v_1 = t)[a]$, while if t is a definable class which is *not* a set we put $t^{(M)} = \{a \in M \colon M \models \varphi[a]\}$.

An $\mathcal{L}$-formula is said to be *restricted* if each of its quantifiers occurs in the form $\forall x \in y$ or $\exists x \in y$ (i.e. $\forall x (x \in y \to \cdots)$ or $\exists x (x \in y \wedge \cdots))$ or if it can be proved equivalent in ZFC to a formula of this kind. The formula $\mathrm{Ord}(x)$ is restricted. A formula is Σ_1 if it is built up from atomic formulas and their negations using only the logical operations $\wedge, \vee, \forall x \in y, \exists x$, or if it can be proved equivalent in ZFC to such a formula. The formula $L(x)$ is Σ_1. The importance of restricted and Σ_1-formulas lies in the following facts. Let $\varphi(v_1, \ldots, v_n)$ be an $\mathcal{L}$-formula, M a transitive $\in$-model of ZFC, and $a_1, \ldots, a_n \in M$. If φ is restricted, then

$$M \models \varphi[a_1, \ldots, a_n] \leftrightarrow \varphi(a_1, \ldots, a_n),$$

while if φ is Σ_1,

$$M \models \varphi[a_1, \ldots, a_n] \rightarrow \varphi(a_1, \ldots, a_n).$$

In conclusion, we point out that, *unless otherwise indicated,* all theorems lemmas, problems, etc. in this book are to be regarded as being *proved in* ZFC.

Construction of the model

Suppose that for each set $x \in V$ we are given a *characteristic function* for x, that is, a function c_x taking values in the Boolean algebra $2 = \{0, 1\}$ such that $x \subseteq \mathrm{dom}(c_x)$ and, for all $y \in \mathrm{dom}(c_x), c_x(y) = 1$ iff $y \in x$. It is clear that all information about x is carried by c_x, so it is natural to *identify* x with c_x. If we perform this identification for all $x \in V$, we see that V may, in a natural sense, be regarded as a class of two-valued functions. The snag here is that, although the process turns each $x \in V$ into a two-valued function, the function fails to be homogeneous in that its *domain* does *not* (in general) consist of two-valued functions.

Let us examine this notion of *homogeneity* a little more closely. It is clear that, however we go about defining it, we should require a two-valued function to be homogeneous iff its domain is a set of homogeneous two-valued functions. Now this looks very much like a definition by *recursion*; and indeed the recursion in question can be explicitly performed as follows. By transfinite recursion on α we define

$$V_\alpha^{(2)} = \{x \colon \mathrm{Fun}(x) \wedge \mathrm{ran}(x) \subseteq 2 \wedge \exists \xi < \alpha[\mathrm{dom}(x) \subseteq V_\xi^{(2)}]\} \qquad (1.1)$$

(compare the definition of the V_α!), and then put

$$V^{(2)} = \{x \colon \exists \alpha[x \in V_\alpha^{(2)}]\}. \qquad (1.2)$$

$V^{(2)}$ is then the required class of all homogeneous two-valued functions, since it is easy to see that we have

$$x \in V^{(2)} \leftrightarrow \mathrm{Fun}(x) \wedge \mathrm{ran}(x) \subseteq 2 \wedge \mathrm{dom}(x) \subseteq V^{(2)}. \qquad (1.3)$$

In future we shall drop the cumbersome term 'homogeneous two-valued function' and call the members of $V^{(2)}$ simply *two-valued sets.* From (1.3) we see that a two-valued set is a two-valued function whose domain is a set of two-valued sets. The class $V^{(2)}$ is called the *universe of two-valued sets*; we shall see later on that, as expected, it is in a natural sense *isomorphic* to the standard universe V of sets.

What we now propose to do is to replace the Boolean algebra 2 by an arbitrary *complete Boolean algebra B*, thus obtaining what we shall call the *universe $V^{(B)}$ of B-valued sets*. We shall show that there is a natural way of assigning to each sentence σ of the language of set theory an element $[\![\sigma]\!]^B$ of B which will act as the 'Boolean truth value' of σ in the universe $V^{(B)}$. Calling the sentence σ *true* in $V^{(B)}$ if $[\![\sigma]\!]^B = 1_B$, and *false* in $V^{(B)}$ if $[\![\sigma]\!]^B = 0_B$ (*cf.* classical notion of truth and falsehood), we show that, for any complete Boolean algebra B, all the theorems of ZFC are true in $V^{(B)}$, or, to put it more suggestively, that $V^{(B)}$ is a 'Boolean-valued model' of ZFC. On the other hand, we shall see that, by selecting B carefully, we can arrange for a variety of set-theoretic assertions, for example, the continuum hypothesis or the axiom of constructibility, to be *false* in $V^{(B)}$. The failure of the continuum hypothesis in $V^{(B)}$ will be achieved by selecting B in such a way that, in $V^{(B)}, \omega$ has many (e.g. $\aleph_2$ or $\aleph_{\omega+1}$) 'B-valued subsets' which are not subsets in the two-valued sense. In this way we will establish the independence of the continuum hypothesis from ZFC.

We now suppose given a complete Boolean algebra B, which we will assume to be fixed throughout the rest of this chapter. We also assume that B is a *set*, that is, $B \in V$.

We define the *universe $V^{(B)}$ of B-valued sets* by analogy with (1.2); namely, we define, by recursion on α,

$$V_\alpha^{(B)} = \{x \colon \mathrm{Fun}(x) \wedge \mathrm{ran}(x) \subseteq B \wedge \exists \xi < \alpha[\mathrm{dom}(x) \subseteq V_\xi^{(B)}]\} \qquad (1.4)$$

and

$$V^{(B)} = \{x \colon \exists \alpha[x \in V_\alpha^{(B)}]\}. \qquad (1.5)$$

We see immediately that, as in (1.3), we have

$$x \in V^{(B)} \leftrightarrow \mathrm{Fun}(x) \wedge \mathrm{ran}(x) \subseteq B \wedge \mathrm{dom}(x) \subseteq V^{(B)}, \qquad (1.6)$$

that is, a *B*-valued set is a *B*-valued function whose domain is a set of *B*-valued sets. $V^{(B)}$ is called a *Boolean extension of V*, or, more precisely, the *B-extension of V*.

An easy induction on rank argument proves:

1.7 Induction Principle for $V^{(B)}$. For *any formula $\phi(x)$,*

$$\forall x \in V^{(B)}[\forall y \in \mathrm{dom}(x)\phi(y) \to \phi(x)] \to \forall x \in V^{(B)}\phi(x).$$

We now introduce a first-order language suitable for making statements about $V^{(B)}$. Let $\mathcal{L}^{(B)}$ be the first-order language obtained from $\mathcal{L}$ by adding a name for each element of $V^{(B)}$. *For convenience we agree to identify each element of $V^{(B)}$ with its name in $\mathcal{L}^{(B)}$.* By coding the formulas of $\mathcal{L}^{(B)}$ as sets in V in the usual way, it is clear that the collection of formulas of $\mathcal{L}^{(B)}$ becomes a definable class.

At this point it will be convenient to introduce the following terminological convention, which will be adhered to throughout the rest of the book. By *formula*, or *sentence* we shall mean $\mathcal{L}$-formula, or $\mathcal{L}$-sentence, respectively, and by *B-formula*, or *B-sentence* we shall mean $\mathcal{L}^{(B)}$-formula, or $\mathcal{L}^{(B)}$-sentence, respectively.

We next set about constructing the map $[\![\cdot]\!]^B$ from the class of all B-sentences to B, which assigns to each B-sentence σ the *Boolean truth value* of σ in $V^{(B)}$.

Suppose for the sake of argument that Boolean truth values have been assigned to all *atomic B-sentences*, that is, sentences of the form $u = v, u \in v$, for $u, v \in V^{(B)}$. Then it is natural—by analogy with the classical two-valued case—to extend the assignment of Boolean truth values to all B-sentences inductively as follows. For B-sentences σ, τ we put

$$[\![\sigma \wedge \tau]\!]^B =_{\mathrm{df}} [\![\sigma]\!]^B \wedge [\![\tau]\!]^B; \tag{1.8}$$

$$[\![\neg\sigma]\!]^B =_{\mathrm{df}} ([\![\sigma]\!]^B)^*. \tag{1.9}$$

If $\phi(x)$ is a B-formula with one free variable x, such that $[\![\phi(u)]\!]^B$ has been defined for all $u \in V^{(B)}$, we observe that the definable class $\{[\![\phi(u)]\!]^B : u \in V^{(B)}\}$ is a sub*set* of B and define

$$[\![\exists x \phi(x)]\!]^B =_{\mathrm{df}} \bigvee_{u \in V^{(B)}} [\![\phi(u)]\!]^B. \tag{1.10}$$

From (1.8)–(1.10) it follows immediately that

$$[\![\sigma \vee \tau]\!]^B = [\![\sigma]\!]^B \vee [\![\tau]\!]^B; \tag{1.11}$$

$$[\![\sigma \rightarrow \tau]\!]^B = [\![\sigma]\!]^B \Rightarrow [\![\tau]\!]^B; \tag{1.12}$$

$$[\![\sigma \leftrightarrow \tau]\!]^B = [\![\sigma]\!]^B \Leftrightarrow [\![\tau]\!]^B; \tag{1.13}$$

$$[\![\forall x \phi(x)]\!]^B = \bigwedge_{u \in V^{(B)}} [\![\phi(u)]\!]^B. \tag{1.14}$$

It remains to assign Boolean truth values to the *atomic B-sentences*. Now we certainly want the *axiom of extensionality* to hold in $V^{(B)}$, so we should have

$$[\![u = v]\!]^B = [\![\forall x \in u[x \in v] \wedge \forall y \in v[y \in u]]\!]^B.$$

Also, in accordance with the *logical* truth $u \in v \leftrightarrow \exists y \in v[u = y]$, which we certainly want to be true in $V^{(B)}$, it should be the case that

$$[\![u \in v]\!]^B = [\![\exists y \in v[u = y]]\!]^B.$$

Finally, we shall require that the Boolean truth value of restricted formulas like $\exists x \in u\, \phi(x)$ and $\forall x \in u\, \phi(x)$ depend only on the Boolean truth values of $\phi(x)$ for those x *which are actually in* $\mathrm{dom}(u)$. Moreover, in evaluating the Boolean truth value of such formulas, we agree to be guided by our original case of *characteristic functions*, where, for $x \in \mathrm{dom}(u)$, the 'truth value' of the formula $x \in u$ is $u(x)$. Granded all this, it seems reasonable to require that

$$[\![\exists x \in u\, \phi(x)]\!]^B = \bigvee_{x \in \mathrm{dom}(u)} [u(x) \wedge [\![\phi(x)]\!]^B]$$

and

$$[\![\forall x \in u\, \phi(x)]\!]^B = \bigwedge_{x \in \mathrm{dom}(u)} [u(x) \Rightarrow [\![\phi(x)]\!]^B].$$

Putting these things together, we see that we must have, for $u, v \in V^{(B)}$,

$$[\![u \in v]\!]^B = \bigvee_{y \in \mathrm{dom}(v)} [v(y) \wedge [\![u = y]\!]^B]; \tag{1.15}$$

$$[\![u = v]\!]^B = \bigwedge_{x \in \mathrm{dom}(u)} [u(x) \Rightarrow [\![x \in v]\!]^B]$$

$$\wedge \bigwedge_{y \in \mathrm{dom}(v)} [v(y) \Rightarrow [\![y \in u]\!]^B]. \tag{1.16}$$

Now (1.15) and (1.16) may (and shall) be regarded as a *definition* of $[\![u \in v]\!]^B$ and $[\![u = v]\!]^B$ by recursion on a certain well-founded relation. To see this, define for $x, y, u, v \in V^B$,

$$\langle x, y \rangle < \langle u, v \rangle \text{ iff either } (x \in \mathrm{dom}(u) \text{ and } y = v) \text{ or } (x = u \text{ and } y \in \mathrm{dom}(v)).$$

Then $<$ is easily seen to be a well-founded relation on the class $V^{(B)} \times V^{(B)} = \{\langle x, y \rangle : x \in V^{(B)} \wedge y \in V^{(B)}\}$. If we now put, for $u, v \in V^{(B)}$,

$$G(\langle u, v \rangle) = \langle [\![u \in v]\!]^B, [\![v \in u]\!]^B, [\![u = v]\!]^B, [\![v = u]\!]^B \rangle,$$

then (1.15) and (1.16) may be written, for some class function F,

$$G(\langle u, v \rangle) = F(\langle u, v, G|\{\langle x, y \rangle : \langle x, y \rangle < \langle u, v \rangle\}\rangle).$$

This constitutes a definition of G by recursion on $<$, and from G we obtain $[\![u \in v]\!]^B, [\![u = v]\!]^B$.

Accordingly, we take (1.15) and (1.16) as a *definition* of $[\![\sigma]\!]^B$ for atomic B-sentences σ, and then define $[\![\sigma]\!]^B$ for *all* B-sentences σ by induction on the complexity of σ in accordance with (1.8)–(1.10).

Remarks **1** The construction of $[\![\sigma]\!]^B$ for arbitrary σ evidently has the form of a *truth definition* for set theory and so cannot be completely formalized within the language of set theory. In fact, although the reader will quickly convince himself that for each specific sentence σ of the language of set theory a specific value for $[\![\sigma]\!]^B$ can be written down within that language, the machinery available in ZFC is not (unless ZFC is inconsistent) strong enough to formalize the construction of the map $\sigma \mapsto [\![\sigma]\!]^B$ *as a function of* σ. More precisely, one can prove in ZFC that the collection of all pairs $\langle \sigma, [\![\sigma]\!]^B \rangle$ is not a definable class. We must therefore think of this map as being defined *metalinguistically*. This tiresome point can be circumvented by starting with a specific model M of ZFC such that $M \in V$ and performing the whole construction of $V^{(B)}, [\![\cdot]\!]^B$ within M; *cf.* Chapter 4.

2 We observe that there is a considerable *duplication of elements* in $V^{(B)}$. For example, if for each $\alpha \in \mathrm{ORD}$ we define $Z_\alpha \in V^{(B)}$ by $Z_\alpha = \{\langle x, 0_B \rangle : x \in V_\alpha^{(B)}\}$, it is easy to verify that $[\![Z_\alpha = \emptyset]\!]^B = 1$ for all α, so that each of the different members of the proper class $\{Z_\alpha : \alpha \in \mathrm{ORD}\}$ 'represents' the empty set in $V^{(B)}$. In fact it is not hard to show that, for each $u \in V^{(B)}$ there is a *proper class* of $v \in V^{(B)}$ such that $[\![u = v]\!]^B = 1$. Accordingly it is helpful to think of the members of $V^{(B)}$ as 'representatives' or 'labels' for sets[1] (or even 'potential' sets), on which (Boolean-valued) equality is defined as an *equivalence relation* with very large equivalence classes. The duplication of elements in $V^{(B)}$ could be avoided by agreeing to *identify* all elements $u, v \in V^{(B)}$ such that $[\![u = v]\!]^B = 1$, but there would be no particular gain for our purposes.

We say that a B-sentence σ is *true* or *holds with probability* 1 in $V^{(B)}$, and frequently write

$$V^{(B)} \models \sigma,$$

if $[\![\sigma]\!]^B = 1$. A B-formula is *true* in $V^{(B)}$ if its universal closure is true in $V^{(B)}$. Finally, a rule of inference is *valid* in $V^{(B)}$ if it preserves the truth of formulas in $V^{(B)}$.

From now on we shall usually (although not always) take the liberty of dropping the sub- or super-script from $[\![\sigma]\!]^B, 0_B, 1_B$.

Our next result is basic.

Theorem 1.17 *All the axioms of the first-order predicate calculus with equality are true in* $V^{(B)}$, *and all its rules of inference are valid in* $V^{(B)}$. *In particular, we have*

(i) $[\![u = u]\!] = 1$;

(ii) $u(x) \leq [\![x \in u]\!]$ *for* $x \in \mathrm{dom}(u)$;

(iii) $[\![u = v]\!] = [\![v = u]\!]$;

(iv) $[\![u = v]\!] \wedge [\![v = w]\!] \leq [\![u = w]\!]$;

---

[1] $V^{(B)}$ may also be thought of as a 'label space' in the terminology of Cohen (1966).

(v) $[\![u = v]\!] \wedge [\![u \in w]\!] \leq [\![v \in w]\!]$;
(vi) $[\![v = w]\!] \wedge [\![u \in v]\!] \leq [\![u \in w]\!]$;
(vii) $[\![u = v]\!] \wedge [\![\phi(u)]\!] \leq [\![\phi(v)]\!]$.

for any B-formula $\phi(x)$.

Proof We sketch proofs of (i)–(vii), leaving the rest to the reader.

(i) We employ the induction principle for $V^{(B)}$. Assume as inductive hypothesis that $[\![x = x]\!] = 1$ for $x \in \mathrm{dom}(u)$. Then for $x \in \mathrm{dom}(u)$ we have

$$(*) \qquad [\![x \in u]\!] = \bigvee_{y \in \mathrm{dom}(u)} u(y) \wedge [\![x = y]\!] \geq u(x) \wedge [\![x = x]\!] = u(x).$$

Therefore $[\![u = u]\!] = \bigwedge_{x \in \mathrm{dom}(u)} [u(x) \Rightarrow [\![x \in u]\!]] = 1$, and the result follows.

(ii) is proved as in $(*)$ above, using (i).

(iii) holds by symmetry.

(iv) is proved using the induction principle for $V^{(B)}$. Assume as inductive hypothesis that

$$\forall v, w \in V^{(B)} [[\![x = v]\!] \wedge [\![v = w]\!] \leq [\![x = w]\!]]$$

for $x \in \mathrm{dom}(u)$. It follows that, for $x \in \mathrm{dom}(u), y \in \mathrm{dom}(v), z \in \mathrm{dom}(w)$, we have

$$[\![x = y]\!] \wedge [\![y = z]\!] \wedge w(z) \leq [\![x = z]\!] \wedge w(z).$$

Taking the supremum over z we get, using the definition of $[\![\cdot \in \cdot]\!]$,

$$[\![x = y]\!] \wedge [\![y \in w]\!] \leq [\![x \in w]\!].$$

But from the definition of $[\![\cdot = \cdot]\!]$ we have

$$[\![v = w]\!] \wedge v(y) \leq [\![y \in w]\!]$$

and also

$$[\![v = w]\!] \wedge [\![x = y]\!] \wedge v(y) \leq [\![x \in w]\!].$$

Now take the supremum over y to get

$$[\![x \in v]\!] \wedge [\![v = w]\!] \leq [\![x \in w]\!].$$

Since

$$[\![u = v]\!] \wedge u(x) \le [\![x \in v]\!],$$

it follows that

$$[\![u = v]\!] \wedge [\![v = w]\!] \wedge u(x) \le [\![x \in w]\!]$$

or

$$[\![u = v]\!] \wedge [\![v = w]\!] \le [u(x) \Rightarrow [\![x \in w]\!]].$$

Hence

$$[\![u = v]\!] \wedge [\![v = w]\!] \le \bigwedge_{x \in \mathrm{dom}(u)} [u(x) \Rightarrow [\![x \in w]\!]], \tag{1}$$

Now, using (iii), the inductive hypothesis implies

$$\forall v, w \in V^{(B)}[[\![w = v]\!] \wedge [\![v = x]\!] \le [\![w = x]\!]],$$

and, using this, an argument similar to that for (1) yields

$$[\![w = v]\!] \wedge [\![v = u]\!] \le \bigwedge_{x \in \mathrm{dom}(w)} [w(z) \Rightarrow [\![z \in u]\!]]. \tag{2}$$

Putting (1) and (2) together gives (iv).

(v) If $z \in \mathrm{dom}(w)$, then (iv) gives

$$[\![u = v]\!] \wedge [\![u = z]\!] \wedge w(z) \le [\![v = z]\!] \wedge w(z).$$

Taking the supremum over z, we get (v).

(vi) If $y \in \mathrm{dom}(v)$, then by definition of $[\![v = w]\!]$ we have

$$[\![v = w]\!] \wedge v(y) \le [\![y \in w]\!]$$

and so, using (v), we get

$$[\![v = w]\!] \wedge [\![u = y]\!] \wedge v(y) \le [\![u \in w]\!].$$

Taking the supremum over y gives (vi).

Finally, (vii) is proved by a straightforward induction on the complexity of ϕ, something we leave to the reader. $\qquad\square$

Remark By analogy with the case of characteristic functions, one might expect *equality* to hold in Theorem 1.17(ii). Although it is easy to show that this is not the case in general (a task we entrust to the reader), it is nonetheless 'almost' the case. In fact, let us call an element $v \in V^{(B)}$ *extensional* if $v(x) = [\![x \in v]\!]$ whenever $x \in \mathrm{dom}(v)$. Then for each $u \in V^{(B)}$ there is an *extensional* $v \in V^{(B)}$ such that $[\![u = v]\!] = 1$. (Simply put $v = \{\langle x, [\![x \in u]\!]\rangle \colon x \in \mathrm{dom}(u)\}$.)

It follows from Theorem 1.17 that all the *theorems* of first-order predicate calculus are true in $V^{(B)}$.

We can now *prove* the laws governing the assignment of Boolean truth values to formulas with restricted quantifiers.

Corollary 1.18 *For any B-formula $\phi(x)$ with one free variable x, and all $u \in V^{(B)}$,*

(i) $\displaystyle [\![\exists x \in u\,\phi(x)]\!] = \bigvee_{x \in \mathrm{dom}(u)} [u(x) \wedge [\![\phi(x)]\!]];$

(ii) $\displaystyle [\![\forall x \in u\,\phi(x)]\!] = \bigwedge_{x \in \mathrm{dom}(u)} [u(x) \Rightarrow [\![\phi(x)]\!]].$

Proof We need only establish (i); (ii) then follows by duality. We have

$$[\![\exists x \in u\,\phi(x)]\!] = [\![\exists x[x \in u \wedge \phi(x)]]\!]$$

$$= \bigvee_{y \in V^{(B)}} [\![y \in u \wedge \phi(y)]\!]$$

$$= \bigvee_{y \in V^{(B)}} \bigvee_{x \in \mathrm{dom}(u)} [[\![x = y]\!] \wedge u(x) \wedge [\![\phi(y)]\!]]$$

$$= \bigvee_{x \in \mathrm{dom}(u)} [u(x) \wedge \bigvee_{y \in V^{(B)}} [\![x = y \wedge \phi(y)]\!]]$$

$$= \bigvee_{x \in \mathrm{dom}(u)} [u(x) \wedge [\![\exists y[x = y \wedge \phi(y)]]\!]]$$

$$= \bigvee_{x \in \mathrm{dom}(u)} [u(x) \wedge [\![\phi(x)]\!]]. \qquad \square$$

Our next result shows precisely how the properties of $V^{(B)}$ can be used to produce relative consistency proofs in set theory. Given a theory T, write $Consis(T)$ for 'T is consistent'. Then we have

Theorem 1.19 *Let T, T' be extensions of ZF such that $Consis(\mathrm{ZF}) \rightarrow Consis(T')$, and suppose that in $\mathcal{L}$ we can define a constant term B such that:*

$(*)$ *$T' \vdash B$ is a complete Boolean algebra and, for each axiom τ of T, we have $T' \vdash [\![\tau]\!]^B = 1_B$.*

Then $Consis(\mathrm{ZF}) \rightarrow Consis(\mathrm{T})$.

Proof If T is inconsistent, then for some axioms $\tau_1, \ldots, \tau_n$ of T we would have, for any sentence σ,

$$\vdash \tau_1 \wedge \cdots \wedge \tau_n \to \sigma \wedge \neg\sigma. \tag{1}$$

Now let B be a complete Boolean algebra satisfying $(*)$. Then

$$T' \vdash [\![\tau_1 \wedge \cdots \wedge \tau_n]\!]^B = 1_B. \tag{2}$$

But (1) gives

$$T' \vdash [\![\tau_1 \wedge \cdots \wedge \tau_n]\!]^B \leq [\![\sigma \wedge \neg\sigma]\!]^B = 0_B,$$

so that, by (2)

$$T' \vdash 1_B \leq 0_B,$$

so T', and hence ZF, would be inconsistent. $\qquad\qquad\square$

Using the standard techniques of arithmetization, Theorem 1.19 can be stated as follows: if T, T' are extensions of ZF such that $(*)$ holds and $Consis(\text{ZF}) \to Consis(T')$ is provable in first-order arithmetic, then $Consis(\text{ZF}) \to Consis(T)$ is also provable in first-order artithmetic. Accordingly, Theorem 1.19 shows that method of Boolean-valued models can furnish purely *finitary* relative consistency proofs.

Remark $V^{(B)}$ may be regarded as a *Boolean-valued structure*. (cf. Ch. 0) Given a complete Boolean algebra B, we here consider a *B-valued structure* to be a triple $\mathcal{S}$ consisting of a class S and two maps

$$[\![\cdot = \cdot]\!]_S, [\![\cdot \in \cdot]\!]_S : S \times S \to B$$

satisfying the analogues of (i), (iii)–(vi) of Theorem 1.17, that is,

$$[\![s = s]\!]_S = 1,$$
$$[\![s = t]\!]_S = [\![t = s]\!]_S,$$
$$[\![s = t]\!]_S \wedge [\![t = u]\!]_S \leq [\![s = u]\!]_S,$$
$$[\![s = t]\!]_S \wedge [\![s \in u]\!]_S \leq [\![t \in u]\!]_S,$$
$$[\![t = u]\!]_S \wedge [\![s \in t]\!]_S \leq [\![s \in u]\!]_S,$$

for all $s, t, u \in S$.

The assignment of Boolean values can then be extended recursively to sentences of the language $\mathcal{L}$ augmented by names for all elements of S as in (1.8)–(1.10), that is,

$$[\![\sigma \wedge \tau]\!]_S = [\![\sigma]\!]_S \wedge [\![\tau]\!]_S,$$
$$[\![\neg\sigma]\!]_S = [\![\sigma]\!]_S^*$$
$$[\![\exists x \phi(x)]\!]_S = \bigvee_{s \in S} [\![\phi(s)]\!]_S,$$

One then show easily by induction on complexity of sentences that the analogue of Theorem 1.17(vii) holds for S, that is, for any $\mathcal{L}$-formula $\phi(x)$ and $s, t \in S$,

$$[\![s = t]\!]_S \wedge [\![\phi(s)]\!]_S \leq [\![\phi(t)]\!]_S.$$

Subalgebras and their models

A complete Boolean algebra B' is said to be a *complete subalgebra* of B if B' is a subalgebra of B and, for any $X \subseteq B'$, $\bigvee X$ and $\bigwedge X$ formed in B' are the same as $\bigvee X$ and $\bigwedge X$, respectively, formed in B. Our next result shows that, if B' is a complete subalgebra of B, then $V^{(B')}$ is, in a natural sense, a *submodel* of $V^{(B)}$.

Theorem 1.20 *Let B' be a complete subalgebra of B. Then*

(i) $V^{(B')} \subseteq V^{(B)}$.

Moreover, for $u, v \in V^{(B')}$,

(ii) $[\![u \in v]\!]^{B'} = [\![u \in v]\!]^{B}$;

(iii) $[\![u = v]\!]^{B'} = [\![u = v]\!]^{B}$.

Proof (i) is clear, while (ii) and (iii) are proved simultaneously by induction on the well-founded relation $y \in \mathrm{dom}(x)$. Details are left to the reader. (*Hint*: the inductive hypothesis is: for all $y \in \mathrm{dom}(v)$ and all $u \in V^{(B)}$,

$$[\![u \in y]\!]^{B'} = [\![u \in y]\!]^{B}$$
$$[\![u = y]\!]^{B'} = [\![u = y]\!]^{B}$$
$$[\![y \in u]\!]^{B'} = [\![y \in u]\!]^{B}.)$$
$\qquad\square$

Corollary 1.21 *If B' is a complete subalgebra of B, then, for any restricted formula $\phi(v_1, \ldots, v_n)$ and any $u_1, \ldots, u_n \in V^{(B')}$,*

$$[\![\phi(u_1, \ldots, u_n)]\!]^{B'} = [\![\phi(u_1, \ldots, u_n)]\!]^{B}.$$

Proof By induction on the complexity of ϕ. For atomic ϕ the result holds by Theorem 1.20. The only nontrivial induction step arises when ϕ is $\exists x \in u\psi$. And here we argue as follows: if $u, u_1, \ldots, u_n \in V^{(B)}$, then, writing $\bigvee^B, \bigvee^{B'}$ for joins in B, B' respectively,

$$\llbracket \phi(u, u_1, \ldots, u_n) \rrbracket^{B'} = \bigvee_{x \in \mathrm{dom}(u)}^{B'} [u(x) \wedge \llbracket \psi(x, u_1, \ldots, u_n) \rrbracket^{B'}]$$

$$= \bigvee_{x \in \mathrm{dom}(u)}^{B} [u(x) \wedge \llbracket \psi(x, u_1, \ldots, u_n) \rrbracket^{B}]$$

$$= \llbracket \phi(u, u_1, \ldots, u_n) \rrbracket^{B}. \qquad \square$$

In this connection we notice that the two-element algebra $2 = \{0, 1\}$ is a complete subalgebra of *every* complete Boolean algebra B, so that $V^{(2)}$ is a *submodel* of every $V^{(B)}$. We are now going to show that $V^{(2)}$ is, in a certain sense, *isomorphic* to the standard universe V. To this end we make the following definition.

Definition 1.22 For each $x \in V$,

$$\hat{x} = \{\langle \hat{y}, 1 \rangle : y \in x\}.$$

This is a definition by recursion on the well-founded relation $y \in x$. Clearly, for each $x \in V, \hat{x} \in V^{(2)} \subseteq V^{(B)}$. Also, by Theorem 1.20, for $x, y \in V$ we have

$$\llbracket \hat{x} \in \hat{y} \rrbracket^{B} = \llbracket \hat{x} \in \hat{y} \rrbracket^{2} \in 2$$

$$\llbracket \hat{x} = \hat{y} \rrbracket = \llbracket \hat{x} = \hat{y} \rrbracket^{2} \in 2.$$

We may regard $\hat{x}$ as being the natural 'representative' in $V^{(B)}$ of each $x \in V$, and accordingly members of $V^{(B)}$ of the form $\hat{x}$ are called *standard*. Our next result establishes the main facts about the standard members of $V^{(B)}$.

Theorem 1.23   (i) *For $x \in V, u \in V^{(B)}$,*

$$\llbracket u \in \hat{x} \rrbracket = \bigvee_{y \in x} \llbracket u = \hat{y} \rrbracket.$$

(ii) *For $x, y \in V$,*

$$x \in y \leftrightarrow V^{(B)} \models \hat{x} \in \hat{y};$$

$$x = y \leftrightarrow V^{(B)} \models \hat{x} = \hat{y}.$$

(iii) *The map $x \mapsto \hat{x}$ is one–one from V into $V^{(2)}$.*

(iv) *For each $u \in V^{(2)}$ there is a unique $x \in V$ such that $V^{(B)} \models u = \hat{x}$.*

(v) *For any formula $\phi(v_1, \ldots, v_n)$ and any $x_1, \ldots, x_n \in V$,*

$$\phi(x_1, \ldots, x_n) \leftrightarrow V^{(2)} \models \phi(\hat{x}_1, \ldots, \hat{x}_n)$$

and if ϕ is restricted then

$$\phi(x_1, \ldots x_n) \leftrightarrow V^{(B)} \models \phi(\hat{x}_1, \ldots, \hat{x}_n).$$

Proof (i) We have

$$\llbracket u \in \hat{x} \rrbracket = \bigvee_{v \in \mathrm{dom}(\hat{x})} [\hat{x}(v) \wedge \llbracket u = v \rrbracket]$$

$$= \bigvee_{y \in x} [\hat{x}(\hat{y}) \wedge \llbracket u = \hat{y} \rrbracket]$$

$$= \bigvee_{y \in x} \llbracket u = \hat{y} \rrbracket.$$

(ii) is established by induction on rank(y), the induction hypothesis being: for all z with rank(z) < rank (y)

$$\forall x [x \in z \leftrightarrow \llbracket \hat{x} \in \hat{z} \rrbracket = 1]$$

$$\forall x [x = z \leftrightarrow \llbracket \hat{x} = \hat{z} \rrbracket = 1]$$

$$\forall x [z \in x \leftrightarrow \llbracket \hat{z} \in \hat{x} \rrbracket = 1].$$

We leave the tedious but straightforward details to the reader.

(iii) follows immediately from (ii).

(iv) The uniqueness of x follows from (ii). The existence of x is proved by induction on the well-founded relation $x \in \mathrm{dom}(u)$. Suppose then that $u \in V^{(2)}$ and

$$\forall x \in \mathrm{dom}(u) \exists y \in V [\llbracket x = \hat{y} \rrbracket = 1].$$

We want to show that, for some $v \in V$, $[\![u = \hat{v}]\!] = 1$. Now

$$[\![u = \hat{v}]\!] = \bigwedge_{x \in \mathrm{dom}(u)} \left[u(x) \implies [\![x \in \hat{v}]\!] \right] \wedge \bigwedge_{y \in v} [\![\hat{y} \in u]\!].$$

So for $[\![u = \hat{v}]\!] = 1$ it is necessary and sufficient that

$$x \in \mathrm{dom}(u) \to u(x) \leq [\![x \in \hat{v}]\!] = \bigvee_{y \in v} [\![x = \hat{y}]\!]; \tag{1}$$

$$y \in v \to 1 = [\![\hat{y} \in u]\!] = \bigvee_{x \in \mathrm{dom}(u)} [u(x) \wedge [\![x = \hat{y}]\!]]. \tag{2}$$

Clearly, in order to satisfy (2) we must take

$$v = \{ y \in V : \exists x \in \mathrm{dom}(u)[u(x) = 1 \wedge [\![x = \hat{y}]\!] = 1] \}.$$

It follows from (ii) and Replacement that $v \in V$, and an application of the inductive hypothesis shows that v satisfies (1). This proves (iv).

The first part of (v) is proved by induction on the complexity of ϕ, using (ii) and (iv). If ϕ is atomic the result holds by (ii). The only nontrivial induction step arises when ϕ in $\exists x \psi$, and this is handled as follows.

Suppose that $x_1, \ldots, x_n \in V$. If

$$[\![\phi(\hat{x}_1, \ldots, \hat{x}_n)]\!]^2 = 1,$$

then

$$\bigvee_{x \in V^{(2)}} [\![\psi(x, \hat{x}_1, \ldots, \hat{x}_n)]\!]^2 = 1$$

so that

$$[\![\psi(x, \hat{x}_1, \ldots, \hat{x}_n)]\!]^2 = 1$$

for some $x \in V^{(2)}$. But, by (iv), for some $y \in V$ we have $[\![x = \hat{y}]\!]^2 = 1$, so that

$$1 = [\![\psi(x, \hat{x}_1, \ldots, \hat{x}_n)]\!]^2 \wedge [\![x = \hat{y}]\!]^2$$

$$\leq [\![\psi(\hat{y}, \hat{x}_1, \ldots, \hat{x}_n)]\!]^2.$$

The inductive hypothesis now gives $\psi(y, x_1, \ldots, x_n)$, which in turn implies $\phi(x_1, \ldots, x_n)$. The converse is similar.

Finally, the second part of (v) follows from the first part and Corollary 1.21.

$\square$

Parts (iii), (iv), and (v) of the above theorem show that the universe $V^{(2)}$ of two-valued sets is, in a natural sense, *isomorphic* to the standard universe V. In particular, it follows from (v) that V and $V^{(2)}$ have the *same true sentences*.

Problem 1.24 (Σ_1-**formulas in** $V^{(B)}$). Let $\phi(v_1, \dots, v_n)$ be a Σ_1-formula, and let $x_1, \dots, x_n \in V$. Show that

$$\phi(x_1, \dots, x_n) \to V^{(B)} \models \phi(\hat{x}_1, \dots, \hat{x}_n).$$

Mixtures and the Maximum Principle

We are now going to formulate a useful general method for constructing elements of $V^{(B)}$.

Given a subset $\{a_1 : i \in I\} \subseteq B$, and a subset $\{u_i : i \in I\} \subseteq V^{(B)}$, we define the *mixture* $\sum_{i \in I} a_i \cdot u_i$ of $\{u_i : i \in I\}$ *with respect to* $\{a_i : i \in I\}$ to be that element $u \in V^{(B)}$ such that

$$\mathrm{dom}(u) = \bigcup_{i \in I} \mathrm{dom}(u_i)$$

and, for $z \in \mathrm{dom}(u)$,

$$u(z) = \bigvee_{i \in I} [a_i \wedge [\![z \in u_i]\!]].$$

If $I = \{0, 1\}$ we write $a_0 \cdot u_0 + a_1 \cdot u_1$ for $\sum_{i \in I} a_i \cdot u_i$: this is called a *two-term mixture*.

A subset $A \subseteq B$ is called an *antichain* in B if $a \wedge b = 0$ for any distinct elements a, b for A. *If an antichain A is given as an indexed set* $\{a_i : i \in I\}$ *we shall always assume that* $a_i \wedge a_j = 0$ *whenever* $i \neq j$ *in I. A partition of unity in B is an antichain A in B such that* $\bigvee A = 1$.

Our next result justifies the use of the term 'mixture' by showing that under certain mild conditions (in particular, when $\{a_i : i \in I\}$ is an antichain) $\sum_{i \in I} a_i \cdot u_i$ behaves as if it were obtained by 'mixing' the B-valued sets $\{u_i : i \in I\}$ together in (at least) the 'proportions' $\{a_i : i \in I\}$.

Mixing Lemma 1.25 *Let* $\{a_i : i \in I\} \subseteq B$, *let* $\{u_i : i \in I\} \subseteq V^{(B)}$ *and put* $\sum_{i \in I} a_i \cdot u_i = u$. *Suppose that, for all* $i, j \in I$,

$$(*) \qquad\qquad a_i \wedge a_j \leq [\![u_i = u_j]\!].$$

Then, for all $i \in I$,

$$a_i \leq [\![u = u_i]\!].$$

In particular, the result holds if $\{a_i : i \in I\}$ *is an antichain.*

Proof We have $[\![u = u_i]\!] = a \wedge b$, where

$$a = \bigwedge_{z \in \mathrm{dom}(u)} [u(z) \Rightarrow [\![z \in u_i]\!]]$$

$$b = \bigwedge_{z \in \mathrm{dom}(u_i)} [u_i(z) \Rightarrow [\![z \in u]\!]].$$

If $z \in \mathrm{dom}(u)$, then

$$a_i \wedge u(z) = \bigvee_{j \in I} a_i \wedge a_j \wedge [\![z \in u_j]\!]$$

$$\leq \bigvee_{j \in I} [\![u_i = u_j]\!] \wedge [\![z \in u_j]\!] \qquad (\mathrm{by}(*))$$

$$\leq [\![z \in u_i]\!],$$

so that $a_i \leq [u(z) \Rightarrow [\![z \in u_i]\!]]$ for any $z \in \mathrm{dom}(u)$, whence $a_i \leq a$. On the other hand, if $z \in \mathrm{dom}(u_i)$, then

$$a_i \wedge u_i(z) \leq a_i \wedge [\![z \in u_i]\!] \leq u(z) \leq [\![z \in u]\!]$$

so that $a_i \leq [u_i(z) \Rightarrow [\![z \in u]\!]]$, whence $a_i \leq b$. Hence $a_i \leq a \wedge b$, and the result follows. $\qquad\square$

Problem 1.26 (Further properties of mixtures) Let $\{a_i : i \in I\}$ be a partition of unity in B.

  (i) Let $\{x_i : i \in I\} \subseteq V$ be such that $x_i \neq x_j$ whenever $i \neq j$. Show that there is $x \in V^{(B)}$ such that $a_i = [\![x = \hat{x}_i]\!]$ for all $i \in I$.

  (ii) Let $\{u_i : i \in I\} \subseteq V^{(B)}$ and suppose that $v \in V^{(B)}$ satisfies $a_i \leq [\![v = u_i]\!]$ for all $i \in I$. Show that $V^{(B)} \models v = \sum_{i \in I} a_i \cdot u_i$.

Recall that in (1.10) we assigned a Boolean truth value to the formula $\exists x \phi(x)$ by putting

$$[\![\exists x \phi(x)]\!] = \bigvee_{u \in V^{(B)}} [\![\phi(u)]\!].$$

We now show, using the Mixing Lemma, that $V^{(B)}$ contains so many members that the supremum on the right side of the above equality is actually *attained* at some element $u \in V^{(B)}$.

Lemma 1.27 (The Maximum Principle) *If $\phi(x)$ is any B-formula, then there is $u \in V^{(B)}$ such that*

$$\llbracket \exists x \phi(x) \rrbracket = \llbracket \phi(u) \rrbracket.$$

In particular, if $V^{(B)} \models \exists x \phi(x)$, then $V^{(B)} \models \phi(u)$ for some $u \in V^{(B)}$.

Proof By (1.10) we have

$$\llbracket \exists x \phi(x) \rrbracket = \bigvee_{u \in V^{(B)}} \llbracket \phi(u) \rrbracket.$$

Since B is a set, so is $\{\llbracket \phi(u) \rrbracket : u \in V^{(B)}\}$ and the axiom of choice implies that there is an ordinal α and a set $\{u_\xi : \xi < \alpha\} \subseteq V^{(B)}$ such that $\{\llbracket \phi(u) \rrbracket : u \in V^{(B)}\} = \{\llbracket \phi(u_\xi) \rrbracket : \xi < \alpha\}$. Accordingly,

$$\llbracket \exists x \phi(x) \rrbracket = \bigvee_{\xi < \alpha} \llbracket \phi(u_\xi) \rrbracket.$$

For each $\xi < \alpha$, put

$$a_\xi = \llbracket \phi(u_\xi) \rrbracket - \bigvee_{\eta < \xi} \llbracket \phi(u_\eta) \rrbracket.$$

Then $\{a_\xi : \xi < \alpha\}$ is an antichain in B and $a_\xi \leq \llbracket \phi(u_\xi) \rrbracket$ for all $\xi < \alpha$. Put $u = \sum_{\xi < \alpha} a_\xi \cdot u_\xi$; then by the Mixing Lemma we have $a_\xi \leq \llbracket u = u_\xi \rrbracket$ for all $\xi < \alpha$. Also, clearly,

$$\llbracket \phi(u) \rrbracket \leq \llbracket \exists x \phi(x) \rrbracket.$$

On the other hand,

$$\llbracket \phi(u) \rrbracket \geq \llbracket u = u_\xi \rrbracket \wedge \llbracket \phi(u_\xi) \rrbracket \geq a_\xi$$

so that

$$\llbracket \phi(u) \rrbracket \geq \bigvee_{\xi < \alpha} a_\xi = \bigvee_{\xi < \alpha} \llbracket \phi(u_\xi) \rrbracket = \llbracket \exists x \phi(x) \rrbracket. \qquad \square$$

Corollary 1.28 *Let $\phi(x)$ be a B-formula such that $V^{(B)} \models \exists x \phi(x)$.*

   (i) *For any $v \in V^{(B)}$ there is a $u \in V^{(B)}$ such that $\llbracket \phi(u) \rrbracket = 1$ and $\llbracket \phi(v) \rrbracket = \llbracket u = v \rrbracket$.*

(ii) *If $\psi(x)$ is a B-formula such that for any $u \in V^{(B)}, V^{(B)} \models \phi(u)$ implies $V^{(B)} \models \psi(u)$, then $V^{(B)} \models \forall x[\phi(x) \to \psi(x)]$.*

Proof (i) Apply the Maximum Principle to obtain $w \in V^{(B)}$ such that $[\![\phi(w)]\!] = 1$, put $b = [\![\phi(v)]\!]$ and $u = b \cdot v + b^* \cdot w$. Then

$$[\![\phi(u)]\!] \geq [\![u = v \wedge \phi(v)]\!] \vee [\![u = w \wedge \phi(w)]\!] \geq b \vee b^* = 1,$$

and $[\![u = v]\!] = [\![u = v]\!] \wedge [\![\phi(u)]\!] \leq [\![\phi(v)]\!]$. Since $[\![u = v]\!] \geq b = [\![\phi(v)]\!]$ by definition of u, the result follows.

(ii) Assume the hypothesis, and let $v \in V^{(B)}$. Using (i), choose $u \in V^{(B)}$ such that $[\![\phi(u)]\!] = 1$ and $[\![\phi(v)]\!] = [\![u = v]\!]$. Then $[\![\psi(u)]\!] = 1$ and

$$[\![\phi(v)]\!] = [\![u = v]\!] = [\![u = v]\!] \wedge [\![\psi(u)]\!] \leq [\![\psi(v)]\!].$$

The result follows. $\qquad\qquad\qquad\qquad\qquad\qquad\qquad\qquad\qquad\qquad\quad$ $\square$

Problem 1.29 (A variant of the Maximum Principle) Without using the axiom of choice, show that, if $V^{(B)} \models \exists! x \phi(x)$, then $V^{(B)} \models \phi(u)$ for some $u \in V^{(B)}$. (Choose a sufficiently large ordinal α such that $1 = [\![\exists x \phi(x)]\!] = \bigvee_{x \in V^{(B)}_\alpha} [\![\phi(x)]\!]$ and define $u \in V^{(B)}$ by $\mathrm{dom}(u) = V^{(B)}_\alpha, u(z) = [\![\exists x[\phi(x) \wedge z \in x]]\!]$.)

Problem 1.30 (The Maximum Principle is equivalent to the axiom of choice)

(i) Let $\{a_i : i \in I\} \subseteq B$ satisfy $\bigvee_{i \in I} a_i = 1$. A partition of unity $\{b_i : i \in I\}$ in B is called a *disjoint refinement* of $\{a_i : i \in I\}$ if $\forall i \in I[b_i \leq a_i]$. Define $u \in V^{(B)}$ by $\mathrm{dom}(u) = \{\hat{i} : i \in I\}, u(\hat{i}) = a_i$ for $i \in I$. Let R be the set of disjoint refinements of $\{a_i : i \in I\}$ and $U = \{v \in V^{(B)} : [\![v \in u]\!] = 1\}$. Show that the map $\{b_i : i \in I\} \mapsto \sum_{i \in I} b_i \cdot \hat{i}$ from R to U is one–one and 'onto' U in the sense that, for any $v \in U$ there is a unique $\{b_i : i \in I\} \in R$ such that $[\![\sum_{i \in I} b_i \cdot \hat{i} = v]\!] = 1$.

(ii) Let Σ_B be the assertion

$$\forall u \in V^{(B)}[[\![u \neq \varnothing]\!] = 1 \to \exists v \in V^{(B)}[[\![v \in u]\!] = 1]]$$

(every nonempty B-valued set has an element) and Π_B the assertion: 'for any set, I, every I-indexed family of elements of B with join 1 has a disjoint refinement'. Show without using the axiom of choice that Σ_B and Π_B are equivalent. (Use (i).) Deduce that the assertions 'Σ_B holds for every complete Boolean algebra B', and 'the Maximum Principle holds in $V^{(B)}$ for every complete Boolean algebra B' are each equivalent to the axiom of choice. (Confine attention to the case in which B is of the form PX for an arbitrary set X.)

We conclude this section by introducing the notion of a core of a Boolean-valued set. Let $u \in V^{(B)}$. A set $v \subseteq V^{(B)}$ is called a *core* for u if the following conditions are satisfied:(i) $[\![x \in u]\!] = 1$ for all $x \in v$, (ii) for each $y \in V^{(B)}$ such that $[\![y \in u]\!] = 1$ there is a *unique* $x \in v$ such that $[\![x = y]\!] = 1$. Thus a core for u represents the class of B -valued objects, which are elements of u with probability 1.

Lemma 1.31 *Any $u \in V^{(B)}$ has a core.*

Proof For each $x \in V^{(B)}$ put

$$f_x = \{\langle z, u(z) \wedge [\![z = x]\!]\rangle : z \in \mathrm{dom}(u)\}.$$

Using the axiom of replacement we can find a set $w \subseteq V^{(B)}$ such that for each $x \in V^{(B)}$ there is $y \in w$ for which $f_x = f_y$. Now let v be a set obtained by selecting one member from each $\sim$-equivalence class in the set $\{x \in w : [\![x \in u]\!] = 1\}$, where $\sim$ is defined by $x \sim y \leftrightarrow [\![x = y]\!] = 1$. It is easily verified that v is a core for u. $\qquad\square$

Note that a core of a B-valued set is unique up to bijection in the sense that there is a bijection between any pair of such cores. Observe also that, if u is a B-valued set the that $V^{(B)} \models u \neq \emptyset$, then the Maximum Principle implies that any core of u is nonempty.

The following result, which will prove useful later on, is an immediate consequence of Corollary 1.28.

Lemma 1.32 *Suppose that $u \in V^{(B)}$ is such that $V^{(B)} \models u \neq \emptyset$ and let v be a core for u. Then for any $x \in V^{(B)}$ there is $y \in v$ such that $[\![x = y]\!] = [\![x \in u]\!]$.*

The truth of the axioms of set theory in $V^{(B)}$

We are now going to show that all of the axioms of ZFC are true in $V^{(B)}$ for any complete Boolean algebra B. (This result is usually expressed by saying that $V^{(B)}$ is a *Boolean valued model* of ZFC.) We do this by *proving* in ZFC that $V^{(B)} \models \sigma$ for each axiom of ZFC.

Theorem 1.33 *All the axioms—and hence all the theorems—of ZFC are true in $V^{(B)}$.*

We prove this theorem by means of a sequence of lemmas.

Lemma 1.34 *The axiom of extensionality is true in $V^{(B)}$.*

Proof This follows immediately from (1.16) and 1.18 (ii). $\qquad\square$

Lemma 1.35 *The axiom scheme of separation is true in $V^{(B)}$.*

Proof Recall that the scheme in question is

$$\forall u \exists v \forall x [x \in v \leftrightarrow x \in u \wedge \psi(x)].$$

To see that each instance is true in $V^{(B)}$, let $u \in V^{(B)}$, define $v \in V^{(B)}$ by $\mathrm{dom}(v) = \mathrm{dom}(u)$ and, for $x \in \mathrm{dom}(v)$,

$$v(x) = u(x) \wedge [\![\psi(x)]\!].$$

Then

$$[\![\forall x [x \in v \leftrightarrow x \in u \wedge \psi(x)]]\!]$$
$$= [\![\forall x \in v [x \in u \wedge \psi(x)]]\!] \wedge [\![\forall x \in u [\psi(x) \to x \in v]]\!].$$

Now

$$[\![\forall x \in v [x \in u \wedge \psi(x)]]\!]$$
$$= \bigwedge_{x \in \mathrm{dom}(v)} [[u(x) \wedge [\![\psi(x)]\!]] \Rightarrow [[\![x \in u]\!] \wedge [\![\psi(x)]\!]]] = 1,$$

using 1.17(ii). Similarly,

$$[\![\forall x \in u [\psi(x) \to x \in v]]\!] = 1,$$

and the assertion follows. $\qquad\qquad\square$

Lemma 1.36 *The axiom scheme of replacement is true in $V^{(B)}$.*

Proof Recall that this is the scheme

$$\forall u [\forall x \in y \exists y \phi(x,y) \to \exists v \forall x \in u \exists y \in v \phi(x,y)].$$

To show that each instance is true in $V^{(B)}$, notice that, for $u \in V^{(B)}$ we have

$$[\![\forall x \in u \exists y \phi(x,y)]\!] = \bigwedge_{x \in \mathrm{dom}(u)} [u(x) \Rightarrow [\![\exists y \phi(x,y)]\!]]$$

$$= \bigwedge_{x \in \mathrm{dom}(u)} \left[u(x) \Rightarrow \bigvee_{y \in V^{(B)}} [\![\phi(x,y)]\!] \right]. \qquad (1)$$

Since B is a set, we may invoke the axiom of replacement in V to obtain a map $x \mapsto \alpha_x$ with domain $\mathrm{dom}(u)$ and range a set of ordinals such that, for each $x \in \mathrm{dom}(u)$,

$$\bigvee_{y \in V^{(B)}} [\![\phi(x,y)]\!] = \bigvee_{y \in V_{\alpha_x}^{(B)}} [\![\phi(x,y)]\!]. \tag{2}$$

Let $\alpha = \bigcup \{\alpha_x : x \in \mathrm{dom}(u)\}$. Then, by (2),

$$\bigwedge_{x \in \mathrm{dom}(u)} \left[u(x) \Rightarrow \bigvee_{y \in V^{(B)}} [\![\phi(x,y)]\!] \right] = \bigwedge_{x \in \mathrm{dom}(u)} \left[u(x) \Rightarrow \bigvee_{y \in V_{\alpha_x}^{(B)}} [\![\phi(x,y)]\!] \right] \tag{3}$$

$$\leq \bigwedge_{x \in \mathrm{dom}(u)} \left[u(x) \Rightarrow \bigvee_{y \in V_{\alpha}^{(B)}} [\![\phi(x,y)]\!] \right].$$

Now put $v = V_{\alpha}^{(B)} \times \{1\}$; then $v \in V^{(B)}$ and

$$\bigvee_{y \in V_{\alpha}^{(B)}} [\![\phi(x,y)]\!] = [\![\exists y \in v\, \phi(x,y)]\!].$$

Hence, by (1) and (3),

$$[\![\forall x \in u\, \exists y\, \phi(x,y)]\!] \leq \bigwedge_{x \in \mathrm{dom}(u)} [u(x) \Rightarrow [\![\exists y \in v\, \phi(x,y)]\!]]$$

$$- [\![\forall x \in u\, \exists y \in v\, \phi(x,y)]\!].$$

The truth of the axiom scheme of replacement in $V^{(B)}$ follows. $\qquad \square$

Lemma 1.37 *The axiom of union is true in* $V^{(B)}$.

Proof This is the sentence

$$\forall u \exists v \forall x [x \in v \leftrightarrow \exists y \in u[x \in y]].$$

To verify its truth in $V^{(B)}$, let $u \in V^{(B)}$; define $v \in V^{(B)}$ so that $\mathrm{dom}(v) = \bigcup \{\mathrm{dom}(y) : y \in \mathrm{dom}(u)\}$ and

$$v(x) = [\![\exists y \in u[x \in y]]\!]$$

for $x \in \mathrm{dom}(v)$. Then

$$\llbracket \forall x \in v \exists y \in u[x \in y] \rrbracket = \bigwedge_{x \in \mathrm{dom}(v)} [\llbracket \exists y \in u[x \in y] \rrbracket \Rightarrow \llbracket \exists y \in u[x \in y] \rrbracket] = 1.$$

Also,

$$\llbracket \forall y \in u \forall x \in y[x \in v] \rrbracket = \bigwedge_{y \in \mathrm{dom}(u)} \left[u(y) \Rightarrow \bigwedge_{x \in \mathrm{dom}(y)} [y(x) \Rightarrow \llbracket x \in v \rrbracket] \right]$$

$$= \bigwedge_{y \in \mathrm{dom}(u)} \bigwedge_{x \in \mathrm{dom}(y)} [u(y) \wedge y(x) \Rightarrow \llbracket x \in v \rrbracket]$$

$$= a, \text{say}.$$

Since $x \in \mathrm{dom}(y)$ and $y \in \mathrm{dom}(u) \to x \in \mathrm{dom}(v)$, we have $\llbracket x \in v \rrbracket \geq v(x)$ for $x \in \mathrm{dom}(y)$. Also, for $x \in \mathrm{dom}(y)$ and $y \in \mathrm{dom}(u)$ we have

$$u(y) \wedge y(x) \leq u(y) \wedge \llbracket x \in y \rrbracket$$

$$\leq \bigvee_{y \in \mathrm{dom}(u)} [u(y) \wedge \llbracket x \in y \rrbracket]$$

$$= \llbracket \exists y \in u[x \in y] \rrbracket$$

$$= v(x).$$

Putting these facts together, we see that

$$a \geq \bigwedge_{y \in \mathrm{dom}(u)} \bigwedge_{x \in \mathrm{dom}(y)} [v(x) \Rightarrow v(x)] = 1$$

and the result follows. $\qquad\square$

Lemma 1.38 *The power set axiom is true in $V^{(B)}$*

Proof This is the sentence

$$\forall u \exists v \forall x[x \in v \leftrightarrow \forall y \in x[y \in u]].$$

To establish its truth in $V^{(B)}$, let $u \in V^{(B)}$ and define $v \in V^{(B)}$ by

$$\mathrm{dom}(v) = B^{\mathrm{dom}(u)}$$

and, for $x \in \text{dom}(v)$,

$$v(x) = [\![x \subseteq u]\!] = [\![\forall y \in x[y \in u]]\!].$$

It suffices to show that

$$[\![\forall x[x \in v \leftrightarrow x \subseteq u]]\!] = 1.$$

First, we note that

$$[\![\forall x \in v[x \subseteq v]]\!] = \bigwedge_{x \in \text{dom}(v)} [v(x) \Rightarrow [\![x \subseteq u]\!]]$$

$$= \bigwedge_{x \in \text{dom}(v)} [v(x) \Rightarrow v(x)]$$

$$= 1.$$

It remains to show that

$$[\![\forall x[x \subseteq u \to x \in v]]\!] = 1. \tag{1}$$

Given $x \in V^{(B)}$, define $x' \in V^{(B)}$ by $\text{dom}(x') = \text{dom}(u)$ and $x'(y) = [\![y \in x]\!]$ for $y \in \text{dom}(x')$. Notice that $x' \in \text{dom}(v)$. We show that

$$[\![x \subseteq u \to x = x']\!] = 1 \tag{2}$$

and

$$[\![x \subseteq u \to x' \in v]\!] = 1, \tag{3}$$

from which it will follow immediately that

$$[\![x \subseteq u \to x \in v]\!] = 1,$$

which yields (1).

We observe that, for any $y \in V^{(B)}$,

$$[\![y \in x']\!] = \bigvee_{z \in \text{dom}(u)} [x'(z) \wedge [\![z = y]\!]]$$

$$= \bigvee_{z \in \text{dom}(u)} [[\![z \in x]\!] \wedge [\![z = y]\!]] \leq [\![y \in x]\!].$$

Therefore

$$[\![x' \subseteq x]\!] = [\![\forall y[y \in x' \to y \in x]]\!] = 1. \tag{4}$$

Next, for any $y \in V^{(B)}$ we have

$$
\begin{aligned}
[\![y \in u \wedge y \in x]\!] &= \bigvee_{z \in \mathrm{dom}(u)} [u(z) \wedge [\![y = z]\!] \wedge [\![y \in x]\!]] \\
&\leq \bigvee_{z \in \mathrm{dom}(u)} [[\![y = z]\!] \wedge [\![z \in x]\!]] \\
&= \bigvee_{z \in \mathrm{dom}(u)} [[\![y = z]\!] \wedge x'(z)] \\
&= [\![y \in x']\!],
\end{aligned}
$$

so that $[\![u \cap x \subseteq x']\!] = 1$. Hence, using this and (4), we get

$$[\![x \subseteq u]\!] \leq [\![u \cap x \subseteq x' \wedge x' \subseteq x \wedge x \subseteq u]\!] \leq [\![x = x']\!],$$

which gives (2).

Finally we prove (3). We have

$$
\begin{aligned}
[\![x \subseteq u]\!] &= [\![\forall y[y \in x \to y \in u]]\!] \\
&= \bigwedge_{y \in V^{(B)}} [[\![y \in x]\!] \Rightarrow [\![y \in u]\!]] \\
&\leq \bigwedge_{y \in \mathrm{dom}(x')} [x'(y) \Rightarrow [\![y \in u]\!]] \\
&= [\![\forall y \in x'[y \in u]]\!] \\
&= [\![x' \subseteq u]\!] = v(x') \leq [\![x' \in v]\!],
\end{aligned}
$$

since $x' \in \mathrm{dom}(v)$. This immediately gives (3), completing the proof.   □

In connection with the proof of Lemma 1.38, we make the following

---

Definition 1.39 For $u \in V^{(B)}$ we define $P^{(B)}(u)$ to be that element $v \in V^{(B)}$ such that $\mathrm{dom}(v) = B^{\mathrm{dom}(u)}$ and, for $x \in \mathrm{dom}(v), v(x) = [\![x \subseteq u]\!]$. $P^{(B)}(u)$ is called the *power set of u in $V^{(B)}$*; the proof of Lemma 1.38 shows that it satisfies

$$V^{(B)} \models P^{(B)}(u) = Pu.$$

Problem 1.40 (Definite sets) An element $u \in V^{(B)}$ is said to be *definite* if $u(x) = 1$ for all $x \in \mathrm{dom}(u)$. Let $u \in V^{(B)}$ be definite, and define $w \in V^{(B)}$ by $w = B^{\mathrm{dom}(u)} \times \{1\}$. Show that

$$[\![\forall x [x \in w \leftrightarrow x \subseteq u]]\!] = 1.$$

(Use Definition 1.39.) Thus, if u is *definite*, the simple object $B^{\mathrm{dom}(u)} \times \{1\}$ also serves as the power set of u in $V^{(B)}$.

Lemma 1.41 *The axiom of infinity is true in $V^{(B)}$.*

Proof The axiom in question is the sentence $\exists u \phi(u)$, where $\phi(u)$ is the formula

$$\emptyset \in u \wedge \forall x \in u \exists y \in u(x \in y).$$

Now $\phi(u)$ is obviously a restricted formula, and we certainly have $\phi(\omega)$. Hence, by Theorem 1.23 (v), we get $[\![\phi(\hat{\omega})]\!] = 1$, and so $[\![\exists u \phi(u)]\!] = 1$. $\qquad\square$

Lemma 1.42 *The axiom of regularity is true in $V^{(B)}$.*

Proof The axiom scheme in question is

$$\forall x [\forall y \in x \phi(y) \rightarrow \phi(x)] \rightarrow \forall x \phi(x).$$

To see that each instance is true in $V^{(B)}$, first put

$$b = [\![\forall x [\forall y \in x \phi(y) \rightarrow \phi(x)]]\!].$$

It now suffices to show that, for any $x \in V^{(B)}$,

$$b \leq [\![\phi(x)]\!].$$

We apply the induction principle for $V^{(B)}$ (1.7). Assume for $y \in \mathrm{dom}(x)$ that $b \leq [\![\phi(y)]\!]$. Then

$$b \leq \bigwedge_{y \in \mathrm{dom}(x)} [\![\phi(y)]\!] \leq \bigwedge_{y \in \mathrm{dom}(x)} [x(y) \Rightarrow [\![\phi(y)]\!]]$$

$$= [\![\forall y \in x\,\phi(y)]\!].$$

But

$$b \leq [\![\![\forall y \in x\,\phi(y)]\!] \Rightarrow [\![\phi(x)]\!]],$$

so that

$$b \leq [\![\![\forall y \in x\,\phi(y)]\!] \Rightarrow [\![\phi(x)]\!]] \wedge [\![\forall y \in x\,\phi(y)]\!] \leq [\![\phi(x)]\!],$$

as required. $\qquad\square$

In order to verify the *axiom of choice* in $V^{(B)}$ it suffices to verify the set-theoretically equivalent principle *Zorn's lemma*. Recall that a partially ordered set is said to be *inductive* if chains (i.e. linearly ordered subsets) in it have upper bounds and Zorn's lemma states that any nonempty inductive partially ordered set has a maximal element.

So finally we prove

Lemma 1.43 *Zorn's lemma, and hence the axiom of choice, is true in $V^{(B)}$.*

Proof By Corollary 1.28(ii), it is enough to show that, for any X, $\leq_X \in V^{(B)}$, if $V^{(B)} \models \langle X, \leq_X \rangle$ *is a nonempty inductive partially ordered set* then $V^{(B)} \models \langle X, \leq_X \rangle$ *has a maximal element*. Suppose then that the antecedent of this implication holds. Let Y be a core for X and define the relation $\leq_Y$ on Y by

$$y \leq_Y y' \leftrightarrow [\![y \leq_X y']\!] = 1$$

for $y, y' \in Y$. It is easy to verify that $\leq_Y$ is a partial ordering on Y; we claim that with this partial ordering Y is inductive. For let C be any chain in Y. It is readily shown that $C' = C \times \{1\} \in V^{(B)}$ satisfies

$$V^{(B)} \models C'\, is\ a\ chain\ in\ X.$$

Accordingly, by the Maximum Principle there is $u \in V^{(B)}$ for which

$$V^{(B)} \models u\ is\ an\ upper\ bound\ for\ C'\, in\ X.$$

Now choose $w \in Y$ such that $[\![w = u]\!] = 1$. Then w is an upper bound for C in Y. For if $x \in C$, then clearly $[\![x \in C']\!] = 1$, whence $[\![x \leq_X u]\!] = 1$ so that $[\![x \leq_X w]\!] = 1$, and $x \leq_Y w$.

Therefore Y is inductive as claimed. By Zorn's lemma in V, Y has a maximal element c. Then $[\![c \in X]\!] = 1$; we claim further that

$$V^{(B)} \models c \text{ is a maximal element of } X. \tag{1}$$

To prove this, take $x \in V^{(B)}$ and apply Lemma 1.32 to obtain $y \in Y$ for which $[\![x \in X]\!] = [\![x = y]\!]$. Then

$$[\![c \leq_X x \wedge x \in X]\!] = [\![c \leq_X x \wedge x = y]\!] \leq [\![c \leq_X y]\!]. \tag{2}$$

Now let $v = y \cdot a + c \cdot a^*$, where $a = [\![c \leq_X y]\!]$. Then $[\![v \in X]\!] = 1$ and so there is $z \in Y$ for which $[\![v = z]\!] = 1$. It is easily shown that $[\![c \leq_X v]\!] = 1$, whence $[\![c \leq_X z]\!] = 1$, and so $c \leq_Y z$. Hence $c = z$ by the maximality of c. Therefore

$$\begin{aligned}
[\![c \leq_X y]\!] = a &\leq [\![y = v]\!] \\
&\leq [\![y = v]\!] \wedge [\![v = z]\!] \\
&\leq [\![y = z]\!] \\
&= [\![y = c]\!],
\end{aligned}$$

and so by (2)

$$\begin{aligned}
[\![c \leq_X x \wedge x \in X]\!] &\leq [\![y = c]\!] \wedge [\![x \in X]\!] \\
&\leq [\![y = c]\!] \wedge [\![x = y]\!] \\
&\leq [\![x = c]\!].
\end{aligned}$$

Thus

$$V^{(B)} \models \forall x \in X [c \leq_X x \rightarrow x = c]$$

that is, (1). This completes the proof. $\square$

The proof of Theorem 1.33 is now complete.

Ordinals and constructible sets in $V^{(B)}$

Since the formula $\mathrm{Ord}(x)$ is restricted, it follows from Theorem 1.23(v) that $[\![\mathrm{Ord}(\hat{\alpha})]\!] = 1$ for every ordinal α. It is natural to call members of $V^{(B)}$ of the form $\hat{\alpha}$ *standard* ordinals in $V^{(B)}$. Our next result relates the property of being an (arbitrary) ordinal in $V^{(B)}$ to that of being a standard ordinal.

Theorem 1.44 *For all* $u \in V^{(B)}$,

$$[\![\mathrm{Ord}(u)]\!] = \bigvee_{\alpha \in \mathrm{ORD}} [\![u = \hat{\alpha}]\!].$$

Proof Since $[\![\mathrm{Ord}(\hat{\alpha})]\!] = 1$, we have

$$[\![u = \hat{\alpha}]\!] = [\![u = \hat{\alpha}]\!] \wedge [\![\mathrm{Ord}(\hat{\alpha})]\!] \leq [\![\mathrm{Ord}(u)]\!].$$

Hence

$$\bigvee_{\alpha \in \mathrm{ORD}} [\![u = \hat{\alpha}]\!] \leq [\![\mathrm{Ord}(u)]\!].$$

To establish the reverse inequality, first observe that, since $[\![\hat{\eta} = \hat{\xi}]\!] = 0$ whenever $\eta \neq \xi$ (by Theorem 1.23(ii)), the map $\xi \mapsto [\![x = \hat{\xi}]\!]$ is one–one from

$$D_x = \{\xi \colon [\![x = \hat{\xi}]\!] \neq 0\}$$

into B whenever $x \in \mathrm{dom}(u)$. Since B is a set, so therefore is D_x. Put

$$D = \bigcup_{x \in \mathrm{dom}(u)} D_x.$$

If α_0 is any ordinal greater than every ordinal in D, we have $[\![\hat{\alpha}_0 = x]\!] = 0$ for any $x \in \mathrm{dom}(u)$. Hence

$$[\![\hat{\alpha}_0 \in u]\!] = \bigvee_{x \in \mathrm{dom}(u)} [u(x) \wedge [\![\hat{\alpha}_0 = x]\!]] = 0.$$

A standard theorem of ZF asserts that $\mathrm{Ord}(u) \wedge \mathrm{Ord}(v) \rightarrow u \in v \vee u = v \vee v \in u$; hence

$$[\![\mathrm{Ord}(u)]\!] \leq [\![u \in \hat{\alpha}_0]\!] \vee [\![u = \hat{\alpha}_0]\!] \vee [\![\hat{\alpha}_0 \in u]\!].$$

Since $[\![\hat{\alpha}_0 \in u]\!] = 0$, it follows that

$$[\![\mathrm{Ord}(u)]\!] \leq [\![u \in \hat{\alpha}_0]\!] \vee [\![u = \hat{\alpha}_0]\!] \leq \bigvee_{\alpha \in \mathrm{ORD}} [\![u = \hat{\alpha}]\!]. \qquad \square$$

Problem 1.45 (Boolean-valued ordinals)

(i) Show that, for any formula $\phi(x)$, $[\![\exists\alpha\phi(\alpha)]\!] = \bigvee_\alpha [\![\phi(\hat\alpha)]\!]$ and $[\![\forall\alpha\phi(\alpha)]\!] = \bigwedge_\alpha [\![\phi(\hat\alpha)]\!]$. Thus, quantifications over ordinals in $V^{(B)}$ can be replaced by suprema and infima in B over *standard* ordinals.

(ii) Show that the following conditions on $u \in V^{(B)}$ are equivalent:

    (a) $[\![\mathrm{Ord}(u)]\!] = 1$;

    (b) there is a set A of ordinals and a partition of unity $\{a_\xi : \xi \in A\}$ in B such that $[\![u = \Sigma_{\xi \in A} a_\xi \cdot \hat\xi]\!] = 1$.

Thus the ordinals of $V^{(B)}$ are precisely the mixtures of standard ordinals. Formulate and prove a similar result for the 'natural numbers in $V^{(B)}$'.

The situation for *constructible sets* in $V^{(B)}$ is similar to that for ordinals. In fact, we have the following theorem.

Theorem 1.46 *For all $u \in V^{(B)}$,*

$$[\![L(u)]\!] = \bigvee_{x \in L} [\![u = \hat x]\!].$$

Proof Let L_α be the αth constructible level. Then, using Problem 1.45, we have

$$[\![L(u)]\!] = [\![\exists\alpha(u \in L_\alpha)]\!] = \bigvee_{\alpha \in \mathrm{ORD}} [\![u \in L_{\hat\alpha}]\!].$$

Now the formula $x = L_\alpha$ is Σ_1 (by a well-known result of set theory) so Problem 1.24 gives

$$x = L_\alpha \rightarrow [\![\hat x = L_{\hat\alpha}]\!] = 1,$$

that is

$$[\![\hat L_\alpha = L_{\hat\alpha}]\!] = 1.$$

Therefore

$$[\![L(u)]\!] = \bigvee_{\alpha \in \mathrm{ORD}} [\![u \in L_{\hat\alpha}]\!]$$

$$= \bigvee_{\alpha \in \mathrm{ORD}} [\![u \in \hat L_\alpha]\!]$$

$$= \bigvee_{\alpha \in \mathrm{ORD}} \bigvee_{x \in L_\alpha} [\![u = \hat{x}]\!]$$

$$= \bigvee_{x \in L} [\![u = \hat{x}]\!],$$

which is the required result. $\square$

It follows immediately from this theorem that, if $V \neq L$, then $[\![V \neq L]\!] = 1$.

Problem 1.47 (Boolean-valued constructible sets) State and prove, for constructible sets, results parallel to those give in Problem 1.45.

Cardinals in $V^{(B)}$

We recall that, for each set x, $|x|$ denotes the *cardinality* of x. Since the formula $|x| = |y|$ is easily seen to be Σ_1, it follows immediately from Problem 1.24 that

$$|x| = |y| \rightarrow V^{(B)} \models |\hat{x}| = |\hat{y}|. \tag{1.48}$$

We shall see later on that the converse does not hold in general.

Theorem 1.49

(i) $V^{(B)} \models \hat{\aleph}_0 = \aleph_0$.

(ii) *For all α,*

$$V^{(B)} \models \hat{\aleph}_\alpha \leq \aleph_{\hat{\alpha}}.$$

Proof (i) The formula $x = \aleph_0$ is restricted, so the result in question follows easily from Theorem 1.23(v).

(ii) is proved by induction on α. For $\alpha = 0$ the result follows immediately from (i). Suppose now that $\alpha > 0$ and

$$V^{(B)} \models \hat{\aleph}_\beta \leq \aleph_{\hat{\beta}} \tag{1}$$

for all $\beta < \alpha$. Then we have

$$\aleph_0 \leq \xi < \aleph_\alpha \rightarrow |\xi| = \aleph_\beta \text{ for some } \beta < \alpha$$

$$\rightarrow V^{(B)} \models |\hat{\xi}| = |\hat{\aleph}_\beta| \quad (\text{by} 1.48)$$

$$\rightarrow V^{(B)} \models |\hat{\xi}| \leq \aleph_{\hat{\beta}} \quad (\text{by } (1))$$

$$\rightarrow V^{(B)} \models |\hat{\xi}| < \aleph_{\hat{\alpha}}$$

$$\rightarrow V^{(B)} \models \hat{\xi} < \aleph_{\hat{\alpha}}.$$

Also,

$$\xi < \aleph_0 \to V^{(B)} \models \hat{\xi} < \hat{\aleph}_0$$
$$\to V^{(B)} \models \hat{\xi} < \aleph_{\hat{\alpha}}.$$

Hence

$$\xi < \aleph_\alpha \to V^{(B)} \models \hat{\xi} < \aleph_{\hat{\alpha}}. \tag{2}$$

Thus

$$[\![\eta < \hat{\aleph}_\alpha]\!] = \bigvee_{\xi < \aleph_\alpha} [\![\eta = \hat{\xi}]\!]$$
$$= \bigvee_{\xi < \aleph_\alpha} [\![\![\eta = \hat{\xi}]\!] \wedge [\![\hat{\xi} < \aleph_{\hat{\alpha}}]\!]\!] \quad (\text{by}(2))$$
$$\leq [\![\eta < \aleph_{\hat{\alpha}}]\!].$$

Therefore

$$V^{(B)} \models \forall \eta [\eta < \hat{\aleph}_\alpha \to \eta < \aleph_{\hat{\alpha}}],$$

whence

$$V^{(B)} \models \hat{\aleph}_\alpha \leq \aleph_{\hat{\alpha}}.$$

This completes the induction step, and the proof. $\qquad\square$

Let $\mathrm{Card}(\alpha)$ be the formula which asserts that α is a cardinal. Then we have

Theorem 1.50

(i) $V^{(B)} \models \mathrm{Card}(\hat{\alpha})$ *for any* $\alpha \leq \omega$.

(ii) *If* $V^{(B)} \models \mathrm{Card}(\hat{\alpha})$, *then* $\mathrm{Card}(\alpha)$.

Proof (i) For $\alpha = \omega$ we already know that $V^{(B)} \models \mathrm{Card}(\hat{\alpha})$ by Theorem 1.49 (i). On the other hand, it is a theorem of ZF that $\forall \alpha [\alpha \in \omega \to \mathrm{Card}(\alpha)]$. Hence

$$V^{(B)} \models \forall \alpha [\alpha \in \omega \to \mathrm{Card}(\alpha)].$$

But $V^{(B)} \models \hat{\omega} = \omega$ by 1.49 (i), so that

$$V^{(B)} \models \forall \alpha [\alpha \in \hat{\omega} \to \mathrm{Card}(\alpha)].$$

Hence $\bigwedge_{\alpha \in \omega} [\![\mathrm{Card}(\hat{\alpha})]\!] = 1$, and (i) follows.

(ii) Notice that $\neg \mathrm{Card}(\alpha)$ is a Σ_1-formula, and apply Problem 1.24. $\qquad\square$

The formula $\mathrm{Card}(x)$ is not Σ_1, so the converse of Theorem 1.50(ii) may fail, that is, the property of being a cardinal is not in general preserved under the passage from V to $V^{(B)}$ (*cf.* Chapter 5). However, there is a simple condition on B which ensures that this property *is* preserved.

A Boolean algebra is said to satisfy the *countable chain condition* (ccc) if every antichain in it is *countable*. (It would seem more reasonable to call this the countable *antichain* condition but the present terminology has the sanction of tradition.)

Complete Boolean algebras satisfying the ccc are readily obtained as follows. Let I be any set and let 2^I have the product topology, where 2 is assigned the discrete topology. Then, as is well-known (e.g. see Kelley 1955, prob. 5 $O(f)$), any family of disjoint open sets in 2^I is countable, so *a fortiori* $\mathrm{RO}(2^I)$ satisfies ccc.

We now show that cardinals behave very well in $V^{(B)}$ when B satisfies ccc.

Theorem 1.51 *Suppose that B satisfies ccc. Then, for any α, and any $x, y \in V$,*

   (i) $\mathrm{Card}(\alpha) \to V^{(B)} \models \mathrm{Card}(\hat{\alpha})$;

   (ii) $V^{(B)} \models \aleph_\alpha = \aleph_{\hat{\alpha}}$;

  (iii) $|x| = |y| \leftrightarrow V^{(B)} \models |\hat{x}| = |\hat{y}|$;

  (iv) $\mathrm{Card}(\alpha) \wedge \alpha$ *is regular* $\to V^{(B)} \models \hat{\alpha}$ *is regular*;

   (v) *if α is an uncountable regular cardinal, and $\xi \in V^{(B)}$ satisfies $[\![\xi < \hat{\alpha}]\!] = 1$, then there is an ordinal $\beta < \alpha$ such that $[\![\xi < \hat{\beta}]\!] = 1$.*

Proof (i) Let α be a cardinal. If $\alpha \leq \omega$ then $V^{(B)} \models \mathrm{Card}(\hat{\alpha})$ by Theorem 1.50(i), so we may suppose that $\alpha > \omega$. To obtain the required conclusion it suffices to show that, for all $f \in V^{(B)}$ and all $\beta < \alpha$,

$$[\![\mathrm{Fun}(f) \wedge \mathrm{dom}(f) = \hat{\beta} \wedge \mathrm{ran}(f) = \hat{\alpha}]\!] = 0.$$

Suppose on the contrary that, for some $f \in V^{(B)}$ and $\beta < \alpha$ we have

$$a = [\![\mathrm{Fun}(f) \wedge \mathrm{dom}(f) = \hat{\beta} \wedge \mathrm{ran}(f) = \hat{\alpha}]\!] \neq 0;$$

then

$$0 \neq a \leq \bigwedge_{\eta < \alpha} \bigvee_{\xi < \beta} [\![f(\hat{\xi}) = \hat{\eta}]\!] \wedge a.$$

It follows that for each $\eta < \alpha$ there is a *least* $\xi_\eta < \beta$ such that

$$[\![f(\hat{\xi}_\eta) = \hat{\eta}]\!] \wedge a \neq 0.$$

Since α is an uncountable cardinal and $\beta < \alpha$ there must exist a $\gamma < \beta$ such that the set

$$X = \{\eta < \alpha : \xi_\eta = \gamma\}$$

is uncountable. It follows immediately that the set

$$\{[\![f(\hat{\gamma}) = \hat{\eta}]\!] \wedge a : \eta \in X\}$$

is an uncountable antichain in B, contradicting ccc. Hence $a = 0$ and (i) follows.

(ii) This goes by induction on α. Assuming that $V^{(B)} \models \aleph_\beta = \aleph_{\hat{\beta}}$ for all $\beta < \alpha$, by Theorem 1.49(ii) it suffices to show that

$$V^{(B)} \models \aleph_{\hat{\alpha}} \leq \hat{\aleph}_\alpha.$$

By (i), we have $V^{(B)} \models \mathrm{Card}(\hat{\aleph}_\alpha)$. Also, if $\beta < \alpha$, then $V^{(B)} \models \hat{\aleph}_\beta < \hat{\aleph}_\alpha$ and by inductive hypothesis $V^{(B)} \models \aleph_\beta = \aleph_{\hat{\beta}}$. Hence $V^{(B)} \models \aleph_{\hat{\beta}} < \hat{\aleph}_\alpha$, so that

$$1 = [\![\mathrm{Card}(\hat{\aleph}_\alpha)]\!] \wedge \bigwedge_{\beta < \alpha} [\![\aleph_{\hat{\beta}} < \hat{\aleph}_\alpha]\!]$$

$$= [\![\mathrm{Card}(\hat{\aleph}_\alpha) \wedge \forall \beta < \hat{\alpha}(\aleph_\beta < \hat{\aleph}_\alpha)]\!]$$

$$\leq [\![\aleph_{\hat{\alpha}} \leq \hat{\aleph}_\alpha]\!],$$

completing the induction step, and proving (ii).

(iii) is an immediate consequence of (ii).

(iv) Let α be a regular cardinal; without loss of generality we may assume $\alpha > \aleph_0$. Suppose that the conclusion is false, that is, $[\![\hat{\alpha} \text{ is not regular}]\!] \neq 0$. Let $\phi(x, y)$ be the statement $\mathrm{Fun}(x) \wedge \mathrm{dom}(x) = y$ and $\mathrm{ran}(x)$ is cofinal in $\hat{\alpha}$. Then

$$0 \neq [\![\hat{\alpha} \text{ is not regular}]\!] = [\![\exists \xi < \hat{\alpha} \exists f \phi(f, \xi)]\!] = \bigvee_{\beta < \alpha} [\![\exists f \phi(f, \hat{\beta})]\!].$$

Hence there is $\beta < \alpha$ such that

$$0 \neq [\![\exists f \phi(f, \hat{\beta})]\!] = a, \text{ say,}$$

and so the Maximum Principle yields an $f \in V^{(B)}$ for which $a = [\![\phi(f, \hat{\beta})]\!]$. Then

$$0 \neq a \leq [\![\mathrm{ran}(f) \text{ is cofinal in } \hat{\alpha}]\!]$$

$$= \bigwedge_{\eta < \alpha} \bigvee_{\xi < \beta} \bigvee_{\mu \geq \eta} [\![f(\hat{\xi}) = \hat{\mu}]\!].$$

It follows that for each $\eta < \alpha$ there are ordinals $\xi_\eta < \beta, \mu_\eta < \alpha$ such that $[\![f(\hat{\xi}_\eta) = \hat{\mu}_\eta]\!] \wedge a \neq 0$. Since α is regular and $\beta < \alpha$ there is $\gamma < \beta$ such that $X = \{\eta < \alpha : \xi_\eta = \gamma\}$ has cardinality α. Then

$$\{[\![f(\hat{\gamma}) = \hat{\mu}_\eta]\!] \wedge a : \eta \in X\}$$

is an antichain of cardinality $\alpha > \aleph_0$ in B, contradicting ccc. (iv) follows.

(v). Assuming the hypotheses, the set $\{[\![\xi = \hat{\eta}]\!] : \eta < \alpha\}$ is an antichain in B and hence the set $X = \{\eta < \alpha : [\![\xi = \hat{\eta}]\!] \neq 0\}$ is countable. Put $\beta = \sup X + 1$; then $\beta < \alpha$ and we have

$$1 = [\![\xi < \hat{\alpha}]\!] = \bigvee_{\eta < \alpha} [\![\xi = \hat{\eta}]\!] = \bigwedge_{\eta \in X} [\![\xi = \hat{\eta}]\!]$$

$$\leq \bigvee_{\eta < \beta} [\![\xi = \hat{\eta}]\!] = [\![\xi < \hat{\beta}]\!]$$

and (v) follows. $\qquad\qquad\qquad\qquad\qquad\qquad\qquad\qquad\qquad\qquad\qquad\square$

Finally, we prove a result which will be useful in estimating cardinalities in $V^{(B)}$. We shall need the notion of *ordered pair* in $V^{(B)}$: for $u, v \in V^{(B)}$ we define

$$\{u\}^{(B)} = \{\langle u, 1\rangle\}$$
$$\{u, v\}^{(B)} = \{u\}^{(B)} \cup \{v\}^{(B)}$$
$$\langle u, v\rangle^{(B)} = \{\{u\}^{(B)}, \{u, v\}^{(B)}\}^{(B)}.$$

It is then easily verified that

$$V^{(B)} \models \forall x \forall y \forall u \forall v [\langle x, y\rangle^{(B)} = \langle u, v\rangle^{(B)} \leftrightarrow x = u \wedge y = v].$$

Lemma 1.52 *For any $u \in V^{(B)}$ we can find $f \in V^{(B)}$ such that*

$$V^{(B)} \models \mathrm{Fun}(f) \wedge \mathrm{dom}(f) = \mathrm{dom}(u)\,\hat{}\, \wedge u \subseteq \mathrm{ran}(f)$$

and hence

$$V^{(B)} \models |u| \leq |\mathrm{dom}(u)\,\hat{}\,|.$$

Proof Define

$$f = \{\langle \hat{z}, z\rangle^{(B)} : z \in \mathrm{dom}(u)\} \times \{1\}.$$

Then it is easily verified that $f \in V^{(B)}$ meets the required conditions; to indicate the idea of the proof we show, for example, that $[\![u \subseteq \mathrm{ran}(f)]\!] = 1$. For we have

$$[\![\exists x \cdot \langle x, y \rangle \in f]\!] = \bigvee_{x \in V^{(B)}} [\![\langle x, y \rangle \in f]\!]$$

$$= \bigvee_{z \in \mathrm{dom}(u)} [\![y = z]\!] \wedge \bigvee_{x \in V^{(B)}} [\![x = \hat{z}]\!]$$

$$= \bigvee_{z \in \mathrm{dom}(u)} [\![y = z]\!]$$

$$\geq \bigvee_{z \in \mathrm{dom}(u)} u(z) \wedge [\![y = z]\!]$$

$$= [\![y \in u]\!].$$

The other conditions are verified similarly. $\qquad \square$

Note that Lemma 1.52 yields an alternative proof that the axiom of choice hold in $V^{(B)}$. For, given $u \in V^{(B)}$, there is by the well-ordering theorem in V an ordinal α and a bijection g of α onto $\mathrm{dom}(u)$. It follows that

$$V^{(B)} \models \hat{g} \text{ is a bijection of the ordinal } \hat{\alpha} \text{ onto } \mathrm{dom}(u) \,\hat{}\,.$$

If $f \in V^{(B)}$ is as specified in Lemma 1.52, then

$$V^{(B)} \models f \circ \hat{g} \text{ is a function with domain } \hat{\alpha} \text{ and range } \supseteq u$$

and so

$$V^{(B)} \models u \text{ is well-orderable.}$$

Since this holds for arbitrary $u \in V^{(B)}$, $V^{(B)} \models \mathrm{AC}$.

Problem 1.53 (The κ-chain condition) Let κ be an infinite cardinal. B is said to satisfy the κ-*chain condition* (κ-cc) if each antichain in B has cardinality $< \kappa$. (Thus the ccc is the $\aleph_1$-cc).

(i) Show that B always satisfies $|B|$-cc. (If B contains an antichain of cardinality κ, then $2^{\kappa} \leq |B|$.)
   Now assume that B satisfies κ-cc. Show that:

(ii) $V^{(B)} \models \mathrm{Card}(\hat{\alpha})$ for any cardinal $\alpha > \kappa$;

(iii) if κ is regular, $V^{(B)} \models \mathrm{Card}(\hat{\kappa})$;

(iv) for each $X \subseteq B$ there is $Y \subseteq X$ such that $|Y| < \kappa$ and $\bigvee X = \bigvee Y$. (Let A be a maximal antichain in the ideal generated by X; show that $\bigvee X = \bigvee A$. For each $a \in A$ there is a finite subset $F_a \subseteq X$ such that $a \leq \bigvee F_a$; show that $Y = \bigcup_{a \in A} F_a$ meets the requirements.)

2

FORCING AND SOME INDEPENDENCE PROOFS

The forcing relation

Let $P = \langle P, \leq \rangle$ be a fixed but arbitrary partially ordered set. (We shall use letters p, q, r, p', q', r' to denote elements of P.) Intuitively, the elements of P are to be thought of as states of information about or *conditions* on a set-theoretic state of affairs and the relation $p \leq q$ is to be understood as asserting 'p *refines* q' or 'the information content of p includes that of q'. Two elements p and q of P are said to be *compatible*—written $\mathrm{Comp}(p, q)$—if there is $r \in P$ such that $r \leq p$ and $r \leq q$. The relation $\mathrm{Comp}(p, q)$ is intended to express the assertion that p and q are mutually consistent conditions. P is said to be *refined* if

$$\forall p, q \in P[q \not\leq p \rightarrow \exists p' \leq q \neg \mathrm{Comp}(p, p')].$$

Thus P is refined if whenever q is not a refinement of p, q has a refinement which is incompatible with p.

For each $p \in P$, put

$$O_p = \{q \in P : q \leq p\}.$$

Then, as is easily verified, the O_p form a base for a topology on P called the (left) *order topology*. We put $\mathrm{RO}(P)$ for the complete Boolean algebra of regular open sets in this topology.

Let us call a subset X of a Boolean algebra B *dense* if $0 \notin X$ and for each $0 \neq b \in B$ there is $x \in X$ such that $x \leq b$.

Lemma 2.1

   (i) *P is refined if and only if $O_p \in \mathrm{RO}(P)$ for all $p \in P$.*

   (ii) *If P is refined, the map $p \mapsto O_p$ is an order-isomorphism of P onto a dense subset of $\mathrm{RO}(P)$.*

Proof (i) It is easily verified that, if P is assigned the order topology, then the interior of the closure of a subset X of P is

$$(\overline{X})^\circ = \{q \in P : \forall p' \leq q \exists r \in X[r \leq p']\}.$$

Hence

$$(\overline{O}_p)^\circ = \{q \in P : \forall p' \leq q \exists r \leq p[r \leq p']\}$$
$$= \{q \in P : \forall p' \leq q \; \mathrm{Comp}(p, p')\}. \tag{1}$$

Now suppose that P is refined. We automatically have $O_p \subseteq (\overline{O}_p)^\circ$ since O_p is open. Conversely, if $q \notin O_p$, then $q \not\leq p$, so since P is refined, there is $p' \leq q$ such that $\neg\mathrm{Comp}(p, p')$ and it follows from (1) that $q \notin (\overline{O}_p)^\circ$. Therefore $O_p = (\overline{O}_p)^\circ$, that is, $O_p \in \mathrm{RO}(P)$. Conversely, if $O_p \in \mathrm{RO}(P)$, then $O_p = (\overline{O}_p)^\circ$, so, using (1), we have $q \not\leq p \to q \notin O_p \to q \notin (\overline{O}_p)^\circ \to \exists p' \leq q \neg\mathrm{Comp}(p, p')$, so P is refined . This proves (i).

(ii) follows easily from (i) and the definition of the order topology on P. $\quad\square$

Corollary 2.2 *P is refined iff it is order-isomorphic to a dense subset of a complete Boolean algebra.*

Proof Necessity follows from Lemma 2.1(ii). Conversely, suppose that P is order-isomorphic to a dense subset of a (complete) Boolean algebra B. Then we may identify P with a dense subset of B. If $p, q \in P$ and $q \not\leq p$, then (in B) $q \wedge p^* \neq 0$, so since P is dense there is $p' \in P$ such that $p' \leq q \wedge p^*$. Thus $p' \leq q$ and it is easy to verify that $\neg\mathrm{Comp}(p, p')$. Therefore P is refined. $\quad\square$

Let us say that a pair $\langle B, e \rangle$ (or simply B) is a *Boolean completion of P* if the following conditions are met:

(i) B is a complete Boolean algebra;

(ii) e is an order-isomorphism of P onto a dense subset of B.

Lemma 2.3 *If $\langle B, e \rangle$ and $\langle B', e' \rangle$ are Boolean completions of P, then there is an isomorphism between B and B' which interchanges $e[P]$ and $e'[P]$.*

Proof We give a sketch, leaving the reader to fill in the details. For each $x \in B$ put $P_x = \{p \in P : e(p) \leq x\}$. Then the density of $e[P]$ in B implies that $x = \bigvee e[P_x]$ for each $x \in B$. The map $f : B \to B'$ defined for $x \in B$ by $f(x) = \bigvee e'[P_x]$ then meets the requirements. $\quad\square$

Corollary 2.2 and Lemma 2.3 imply that *each refined partially ordered set P has a Boolean completion which is unique up to isomorphism.*

If P is refined and $\langle B, e \rangle$ is a Boolean completion of P, P will be called a *basis* or *set of conditions* for B (with respect to e). Under these conditions, we shall frequently *identify P with its image $e[P]$ in B*, so that *P will be regarded as a dense subset of B.*

Remark The notion of the Boolean completion of a partially ordered set is closely related to the—possibly more familiar—notion of the completion of a Boolean algebra. A (minimal) completion of a Boolean algebra A is a pair $\langle B, f \rangle$

in which B is a complete Boolean algebra and f is a complete monomorphism of A into B such that $f[A - \{0\}]$ is dense in B.

One easily shows (as in Lemma 2.3) that if $\langle B, f \rangle$ and $\langle B', f' \rangle$ are completions of A, then there is an isomorphism between B and B', which interchanges $f[A]$ and $f'[A]$. We can obtain the completion $\langle B, f \rangle$ of A in either of the following two equivalent ways: (1) take $\langle B, e \rangle$ to be a Boolean completion of the partially ordered set A—$\{0\}$ and define $f : A \to B$ by $f(x) = e(x)$ if $x \neq 0$ and $f(0) = 0$; (2) take B to be the regular open algebra of the Stone space of A and f the natural monomorphism of A into B. The completion $\langle B, f \rangle$ of A is characterized by the following universal property: for any complete Boolean algebra C and any complete homomorphism $g : A \to C$, there is a unique complete homomorphism $h : B \to C$ such that $g = h \circ f$.

Problem 2.4 (Boolean completions of nonrefined sets) Let $\langle P, \leq \rangle$ be a partially ordered set.

(i) Show that there is a refined partially ordered set $\langle Q, \preceq \rangle$ and an order preserving map j of P onto Q such that, for any, $p, q \in P, \mathrm{Comp}(p, q) \leftrightarrow \mathrm{Comp}(jp, jq)$. (Define the equivalence relation $\sim$ on P by $p \sim q \leftrightarrow \forall x[\mathrm{Comp}(p, x) \leftrightarrow \mathrm{Comp}(q, x)]$, and take $Q = P/\sim$.)

(ii) Show that $\langle Q, \preceq \rangle$ is uniquely determined up to isomorphism.

The partially ordered set Q is called the *refined associate* of P and the map j is called the *canonical* map. If P is refined, we may take $Q = P$ and j to be the identity.

(iii) Let B be the Boolean completion of Q; then Q may be identified with a dense subset of B and the canonical map j may be regarded as carrying P into B. Show that j is an order-preserving map onto a dense subset of B such that, for $p, q \in P, \mathrm{Comp}(p, q) \leftrightarrow j(p) \wedge j(q) \neq 0$.

The algebra B is called the *Boolean completion* of P.

Now let x and y be nonempty sets, where y has at least two elements. We put $C(x, y)$ for the set of all mappings with domain a *finite* subset of x and range a subset of y. We agree to partially order $C(x, y)$ by $\supseteq$, that is, *reverse* inclusion, and it is easy to verify that this turns $C(x, y)$ into a *refined* partially ordered set. For $p \in C(x, y)$ we put

$$N(p) = \{f \in y^x : p \subseteq f\},$$

where y^x is, as usual, the set of all mappings from x into y. Subsets of y^x of the form $N(p)$ form a base for the *product topology* on y^x, when y is assigned the discrete topology. Each $N(p)$ is then a *clopen* (i.e. closed-and-open) set in this topology. Thus, in particular, each $N(p)$ is a *regular open* subset of y^x, and it is easy to verify that the map $p \mapsto N(p)$ is an order-isomorphism of $C(x, y)$ onto a dense subset of $\mathrm{RO}(y^x)$. Therefore $\langle \mathrm{RO}(y^x), N \rangle$ is a *Boolean completion* of $C(x, y)$, and the latter is (up to isomorphism) a *basis* for $\mathrm{RO}(y^x)$.

Remark In Cohen's original development of forcing, the ordering of the set of forcing conditions P 'goes up', rather than 'down' as we have taken it. In other words, Cohen holds $p \leq q$ to mean that q contains *more* information than p. In particular, the set $P = C(x, y)$ would be taken to be ordered by *inclusion* and not, as we have stipulated, by inverse inclusion. This (the 'more means more' convention) is of course eminently reasonable, but unfortunately P would then, in a natural sense, be *anti*-refined rather than refined, and one could only show P is order *anti*-isomorphic to a dense subset (or, equivalently, isomorphic to an *anti*-dense subset) of a complete Boolean algebra. (To obtain the 'anti'-version of an order-theoretic concept, interchange '$\leq$' and '$\geq$' and '0' and '1'.) Since anti-isomorphisms and anti-dense subsets are not very convenient technically, we have chosen to reverse the more usual ordering of P and thereby adopt the 'more means less' convention, thus enabling us to use the more familiar machinery of isomorphisms and dense subsets.

Now let B be a complete Boolean algebra, and let P be a basis for B, with respect to an order isomorphism e. Identify P with $e[P]$, so that P becomes a dense subset of B. For each B-sentence σ and each $p \in P$ we define the relation p *forces* σ—written $p \Vdash \sigma$—by

$$p \Vdash \sigma \text{ iff } p \leq \llbracket \sigma \rrbracket^B.$$

The basic properties of the forcing relation are contained in the final theorem of this section. We write $\llbracket \sigma \rrbracket$ for $\llbracket \sigma \rrbracket^B$ as usual.

Theorem 2.5 *let σ and τ be B-sentence and let $\phi(x)$ be a B-formula. Then:*

(i) $p \Vdash \neg\sigma$ *iff* $\neg\exists q \leq p[q \Vdash \sigma]$;

(ii) $p \Vdash \sigma \wedge \tau$ *iff* $p \Vdash \sigma$ *and* $p \Vdash \tau$;

(iii) $p \Vdash \sigma \vee \tau$ *iff* $\forall q \leq p \exists r \leq q[r \Vdash \sigma \text{ or } r \Vdash \tau]$;

(iv) $p \Vdash \sigma \rightarrow \tau$ *iff* $\forall q \leq p[q \Vdash \sigma \rightarrow q \Vdash \tau]$;

(v) $p \Vdash \forall x \phi(x)$ *iff* $\forall u \in V^{(B)} [p \Vdash \phi(u)]$;

(vi) $p \Vdash \exists x \phi(x)$ *iff* $\forall q \leq p \exists r \leq q \exists u \in V^{(B)} [r \Vdash \phi(u)]$;

(vii) *for* $a \in V$, $p \Vdash \forall x \in \hat{a}\phi(x)$ *iff* $\forall x \in a[p \Vdash \phi(\hat{x})]$;

(viii) *for* $a \in V$, $p \Vdash \exists x \in \hat{a}\phi(x)$ *iff* $\forall q \leq p \exists r \leq q \exists x \in a[r \Vdash \phi(\hat{x})]$;

(ix) $\llbracket \sigma \rrbracket = 0$ *iff* $\neg\exists p[p \Vdash \sigma]$;

(x) $\llbracket \sigma \rrbracket = 1$ *iff* $\forall p[p \Vdash \sigma]$;

(xi) $\forall p \exists q \leq p[q \Vdash \sigma \text{ or } q \Vdash \neg\sigma]$;

(xii) $[p \Vdash \sigma] \rightarrow \neg[p \Vdash \neg\sigma]$;

(xiii) $[q \leq p \text{ and } p \Vdash \sigma] \rightarrow q \Vdash \sigma$.

Proof We prove some of these assertions, leaving the rest to the reader.

(i) If $p \Vdash \neg\sigma$, then $p \leq [\![\sigma]\!]^*$. So in this case if $q \leq p$, then $q \not\leq [\![\sigma]\!]$ (otherwise $q \leq [\![\sigma]\!]^* \wedge [\![\sigma]\!] = 0$), so $\neg[q \Vdash \sigma]$. Conversely, if $\neg[p \Vdash \neg\sigma]$, then $p \not\leq [\![\sigma]\!]^*$, so $p \wedge [\![\sigma]\!] \neq 0$, so, since P is dense, $\exists q \leq p[q \leq [\![\sigma]\!]]$, whence $\exists q \leq p[q \Vdash \sigma]$.

(ii)

$$p \Vdash \sigma \wedge \tau \text{ iff } p \leq [\![\sigma]\!] \wedge [\![\tau]\!]$$
$$\text{iff } p \leq [\![\sigma]\!] \text{ and } p \leq [\![\tau]\!]$$
$$\text{iff } p \Vdash \sigma \text{ and } p \Vdash \tau.$$

(iii)

$$p \Vdash \sigma \vee \tau \text{ iff } p \Vdash \neg(\neg\sigma \wedge \neg\tau)$$
$$\text{iff } \neg \, \exists q \leq p[q \Vdash \neg\sigma \wedge \neg\tau]$$
$$\text{iff } \neg\exists q \leq p[q \Vdash \neg\sigma \text{ and } q \Vdash \neg\tau]$$
$$\text{iff } \neg \, \exists q \leq p[\neg \, \exists r \leq q[r \Vdash \sigma] \text{ and } \neg \, \exists r \leq q[r \Vdash \tau]]$$
$$\text{iff } \forall q \leq p[\exists r \leq q[r \Vdash \sigma] \text{ or } \exists r \leq q[r \Vdash \tau]]$$
$$\text{iff } \forall q \leq p\exists r \leq q[r \Vdash \sigma \text{ or } r \Vdash \tau].$$

(vi)

$$p \Vdash \exists x\phi(x) \text{ iff } p \Vdash \neg \, \forall x \, \neg \, \phi(x)$$
$$\text{iff } \neg \, \exists q \leq p[q \Vdash \forall x \, \neg \, \phi(x)]$$
$$\text{iff } \neg\exists q \leq p\forall u \in V^{(B)} \, \neg\exists r \leq q[r \Vdash \phi(u)]$$
$$\text{iff } \forall q \leq p\exists u \in V^{(B)} \, \exists r \leq q[r \Vdash \phi(u)].$$

(xi) Either $p \Vdash \sigma$ or $\neg[p \Vdash \sigma]$. If the former, we are finished. If the latter, then $p \not\leq [\![\sigma]\!]$, so $p \wedge [\![\sigma]\!]^* \neq 0$, whence $\exists q[q \leq p \wedge [\![\sigma]\!]^*]$, so $\exists q \leq p[q \Vdash \neg\sigma]$. $\square$

The meaning of Theorem 2.5 can be clarified as follows. We recall that the elements of p are to be thought of as 'states of information', or briefly, 'states'. Also, for $p, q \in P$, the relation $p \leq q$ means that 'state' p is a refinement (of the information in) 'state' q. Then $p \Vdash \sigma$ may be thought of as asserting that in 'state' p we are in definite possession of the 'fact' σ, or have been 'forced' to accept σ as true. Using this approach, the interpretation of, for example,

(i) in Theorem 2.5 is:

$$p \Vdash \neg\sigma \text{ iff } \textit{in no state which refines } p$$
$$\textit{will we be forced to accept } \sigma;$$

that of (vi) is:

$$p \Vdash \exists x \phi(x) \text{ iff } \textit{any refinement of } p \textit{ can be itself refined}$$
$$\textit{to a state in which we can instantiate } \phi(x);$$

and that of (xi) is:

    each state can be refined to one in which we either accept σ, or we accept $\neg\sigma$.

The reader may provide similar interpretations of the other clauses in Theorem 2.5.

Finally, we point out that the notion of forcing introduced here is what Cohen (1966) called *weak* forcing; Cohen's original notion of forcing (which we shall write $\Vdash_c$) is customarily known as *strong* forcing. The two notions are related by the equivalence

$$p \Vdash \sigma \leftrightarrow p \Vdash_c \neg\neg\sigma. \tag{$*$}$$

The chief difference between weak and strong forcing is that, while the former obeys all the laws of *classical* logic, the latter obeys only the laws of *intuitionistic* logic. This means, for example, that if $p \Vdash \sigma$ then $p \Vdash \tau$ whenever τ is *classically* equivalent to σ, while if $p \Vdash_c \sigma$, then one can only infer that $p \Vdash_c \tau$ when τ satisfies the stronger condition of being *intuitionistically* equivalent to σ. (Note in this connection the resemblance between $(*)$ and the usual translation of classical into intuitionistic logic.)

Independence of the axiom of constructibility and the continuum hypothesis

We are now in a position to use the techniques introduced in earlier sections to prove (among other things) the independence of the axiom of constructibility and the continuum hypothesis from ZFC.

Theorem 2.6 *Let $B = \mathrm{RO}(2^\omega)$. Then:*

  (i) $V^{(B)} \models (P\omega)\hat{\ } \neq P\hat{\omega}.$
  (ii) $V^{(B)} \models P\hat{\omega} \nsubseteq L.$

Proof Put $P = C(\omega, 2)$; then B is a completion of P, P is a basis for B, and each $p \in P$ is identified with the element

$$N(p) = \{f \in 2^\omega : p \subseteq f\}$$

of B. Define $u \in V^{(B)}$ by $\mathrm{dom}(u) = \mathrm{dom}(\hat\omega)$ and

$$u(\hat{n}) = \{f \in 2^\omega : f(n) = 1\} \in B.$$

It is easy to verify that, for $p \in P$ and $n \in \omega$, we have

$$p \Vdash \hat{n} \in u \text{ iff } p(n) = 1;$$
$$p \Vdash \hat{n} \notin u \text{ iff } p(n) = 0.$$

Also,

$$[\![u \in P\hat\omega]\!]^B = [\![u \subseteq \hat\omega]\!]^B = \bigwedge_{n \in \omega} [u(\hat{n}) \Rightarrow [\![\hat{n} \in \hat\omega]\!]^B] = 1.$$

We claim that $[\![u = \hat{x}]\!]^B = 0$ for all $x \in P\omega$. For suppose not; then there is $p \in P$ and $x \in P\omega$ such that $p \Vdash u = \hat{x}$. Pick $n \in \omega$ such that $n \notin \mathrm{dom}(p)$. (Possible, since $\mathrm{dom}(p)$ is finite.) If $n \in x$, put $p' = p \cup \{\langle n, 0 \rangle\}$; if $n \notin x$, put $p' = p \cup \{\langle n, 1 \rangle\}$. Then, if $n \in x$, we have $p' \Vdash \hat{n} \in \hat{x} \wedge \hat{n} \notin u$, and if $n \notin x$, we have $p' \Vdash \hat{n} \notin \hat{x} \wedge \hat{n} \in u$. Thus in either case $p' \Vdash u \neq \hat{x}$. But since $p' \leq p$, we have $p' \Vdash u = \hat{x}$, which is a contradiction. This establishes the claim.

It follows that

$$[\![u \in (P\omega)\hat{\ }]\!]^B = \bigvee_{x \in P\omega} [\![u = \hat{x}]\!]^B = 0,$$

so

$$1 = [\![u \in P\hat\omega]\!]^B \wedge [\![u \notin (P\omega)\hat{\ }]\!]^B \leq [\![P\hat\omega \neq (P\omega)\hat{\ }]\!]^B$$

and (i) is proved.

(ii) Consider the set $u \in V^{(B)}$ defined in the proof of (i). We have, by Theorem 1.46,

$$[\![L(u)]\!]^B = \bigvee_{x \in L} [\![u = \hat{x}]\!]^B$$

$$= \bigvee_{x \in L \cap P\omega} [\![u = \hat{x}]\!]^B \vee \bigvee_{x \in L - P\omega} [\![u = \hat{x}]\!]^B.$$

Since we already know that $[\![u \in P\hat{\omega}]\!]^B = 1$, we have for $x \notin P\omega$,

$$[\![u = \hat{x}]\!]^B = [\![u = \hat{x}]\!]^B \wedge [\![u \in P\hat{\omega}]\!]^B \leq [\![\hat{x} \in P\hat{\omega}]\!]^B = [\![\hat{x} \subseteq \hat{\omega}]\!]^B = 0$$

since $x \nsubseteq \omega$. Therefore

$$[\![L(u)]\!]^B = \bigvee_{x \in L \cap P\omega} [\![u = \hat{x}]\!]^B \leq \bigvee_{x \in P\omega} [\![u = \hat{x}]\!]^B = [\![u \in (P\omega)\hat{\ }]\!]^B = 0.$$

Hence $[\![L(u)]\!]^B = 0$ and (ii) follows immediately. $\qquad\square$

Recall that in the course of proving Theorem 2.6(i) we remarked that, for $p \in P = C(\omega, 2)$, we have $p \Vdash \hat{n} \in u$ iff $p(n) = 1$, $p \Vdash \hat{n} \notin u$ iff $p(n) = 0$. Accordingly, each condition $p \in P$ may be regarded as encoding a finite 'piece of information' about the members of the 'new subset' u of ω. We may therefore think of $C(\omega, 2)$ as *the set of conditions for adjoining a new subset of ω using finite pieces of information,* and its completion $\mathrm{RO}(2^\omega)$ *as an algebra which adjoins a new subset of ω.*

Theorems 1.19, 1.33, and 2.6 now give:

Corollary 2.7 *If* ZF *is consistent, so is* ZFC+ *'there is a nonconstructible subset of ω'.*

We next show how to extend this result to include the GCH.

Theorem 2.8 *Assume the* GCH. *Then, if B satisfies ccc and $|B| = 2^{\aleph_0}$,*

$$V^{(B)} \models \mathrm{GCH}.$$

Proof Recall from Definition 1.39 that we have, for any $u \in V^{(B)}$,

$$\mathrm{dom}(P^{(B)}(u)) = B^{\mathrm{dom}(u)}$$

and

$$V^{(B)} \models P^{(B)}(u) = Pu.$$

Take $u = \hat{\aleph}_\alpha$. Then since the map $x \mapsto \hat{x}$ is one–one, we have $|\mathrm{dom}(\hat{\aleph}_\alpha)| = \aleph_\alpha$. Thus, since the GCH is assumed to hold,

$$|\mathrm{dom}(P^{(B)}(\hat{\aleph}_\alpha))| = |B^{\mathrm{dom}(\hat{\aleph}_\alpha)}| = (2^{\aleph_0})^{\aleph_\alpha} = 2^{\aleph_\alpha} = \aleph_{\alpha+1}.$$

It follows from Lemma 1.52 that

$$V^{(B)} \models |P^{(B)}(\hat{\aleph}_\alpha)| \leq |\hat{\aleph}_{\alpha+1}|.$$

But B satisfies ccc, so, by Theorem 1.51, for any α,

$$V^{(B)} \models \hat{\aleph}_\alpha = \aleph_{\hat{\alpha}}$$

whence

$$V^{(B)} \models |\hat{\aleph}_{\alpha+1}| = \aleph_{\hat{\alpha}+1}.$$

These two facts give

$$V^{(B)} \models |P^{(B)}(\aleph_{\hat{\alpha}})| \leq \aleph_{\hat{\alpha}+1},$$

so that

$$V^{(B)} \models |P\aleph_{\hat{\alpha}}| \leq \aleph_{\hat{\alpha}+1}.$$

Since this holds for *arbitrary* α, it follows using Problem 1.45 that

$$V^{(B)} \models \forall \alpha[|P\aleph_\alpha| \leq \aleph_{\alpha+1}]$$

and so

$$V^{(B)} \models \mathrm{GCH}. \qquad \square$$

Corollary 2.9 *If* ZF *is consistent, so is* ZFC + GCH + *'there is a noncon-structible subset of* ω*'.*

Proof Let B be the algebra introduced in Theorem 2.6. Then B statisfies ccc and $|B| = 2^{\aleph_0}$. Since $Consis(\mathrm{ZF}) \to Consis(\mathrm{ZFC} + \mathrm{GCH})$, the required result now follows immediately from Theorems 2.6, 2.8, 1.33, and 1.19. $\qquad \square$

We next turn to the problem of violating the continuum hypothesis in $V^{(B)}$. The idea here is to make $\mathrm{P}^{(B)}(\hat{\omega})$ large in $V^{(B)}$; we shall see that this can be achieved by taking an appropriate B of large cardinality. On the other hand, if we want to pin down the cardinality of $P^{(B)}(\hat{\omega})$ in $V^{(B)}$, we shall need to make a reasonably precise estimate of $|B|$. We now set about doing this for the sort of B we have in mind.

A topological space X is said to satisfy the *countable chain condition* (ccc) if each disjoint family of sets open in X is countable. We have already remarked that the product space 2^I satisfies ccc.

Lemma 2.10 *Let X be a topological space satisfying ccc. Let E be a base for X and let B be the regular open algebra of X. Then $|B| \leq |E|^{\aleph_0}$.*

Proof Let $U \in B$, and, using Zorn's lemma, let F be a maximal disjoint subfamily of $E \cap PU$. Put $G = \bigcup F$. We claim that $U = (\bar{G})^\circ$. For since $G \subseteq U$ and U is regular open, we have $(\bar{G})^\circ \subseteq (\bar{U})^\circ = U$. On the other hand, consider $U - \bar{G}$. This is an open set; if it is nonempty then it includes a nonempty member of E, which is disjoint from every member of F, contradicting the maximality of F. Thus $U - \bar{G} = \emptyset$, so that $U \subseteq \bar{G}$ and $U \subseteq (\bar{G})^\circ$. This proves the claim.

Accordingly each member of B is determined by a disjoint subfamily of E; since X satisfies ccc each such subfamily is countable and there are at most $|E|^{\aleph_0}$ of them. $\square$

Corollary 2.11 *For each set I let 2^I be the product space where 2 is assigned the discrete topology. If $|I| = \aleph_\alpha$, then*

$$\aleph_\alpha \leq |\mathrm{RO}(2^I)| \leq \aleph_\alpha^{\aleph_0}.$$

Proof The family of sets of the form

$$\{f \in 2^I : f(i_1) = a_1, \ldots, f(i_n) = a_n\},$$

where $i_1, \ldots, i_n \in I$ and $a_1, \ldots, a_n \in 2$ is a base for 2^I of cardinality $\aleph_\alpha$. Since each set of this form is clopen, it is in $\mathrm{RO}(2^I)$ and so $\aleph_\alpha \leq |\mathrm{RO}(2^I)|$. On the other hand, 2^I satisfies ccc and so Lemma 2.10 applies to yield the other inequality. $\square$

We are now in a position to prove

Theorem 2.12 *Suppose that $\aleph_\alpha^{\aleph_0} = \aleph_\alpha$ and let $B = \mathrm{RO}(2^{\omega \times \omega_\alpha})$. Then*

$$V^{(B)} \models 2^{\aleph_0} = \aleph_{\hat{\alpha}}.$$

Proof By Corollary 2.11 we have

$$\aleph_\alpha \leq |B| \leq \aleph_\alpha^{\aleph_0} = \aleph_\alpha,$$

so that $|B| = \aleph_\alpha$. Hence

$$|\mathrm{dom}(P^{(B)}(\hat{\omega}))| = |B^{\mathrm{dom}(\hat{\omega})}| = \aleph_\alpha^{\aleph_0} = \aleph_\alpha$$

and so, by Lemma 1.52

$$V^{(B)} \models |P^{(B)}(\hat{\omega})| \leq |\hat{\aleph}_\alpha|,$$

whence

$$V^{(B)} \models |P\omega| \leq |\hat{\aleph}_\alpha|.$$

But B satisfies ccc, so by Theorem 1.51 we have $V^{(B)} \models |\hat{\aleph}_\alpha| = \aleph_{\hat{\alpha}}$, whence

$$V^{(B)} \models |P_\omega| \leq \aleph_{\hat{\alpha}},$$

that is

$$V^{(B)} \models 2^{\aleph_0} \leq \aleph_{\hat{\alpha}}.$$

It remains to show that

$$V^{(B)} \models \aleph_{\hat{\alpha}} \leq 2^{\aleph_0}.$$

To this end, for each $\nu < \omega_\alpha$ define $u_\nu \in V^{(B)}$ by $\mathrm{dom}(u_\nu) = \mathrm{dom}(\hat{\omega})$ and

$$u_\nu(\hat{n}) = \{ f \in 2^{\omega \times \omega_\alpha} : f(n, \nu) = 1 \}.$$

We have

$$[\![u_\nu \subseteq \hat{\omega}]\!]^B = \bigwedge_{n \in \omega} [\![u_\nu(\hat{n}) \Rightarrow [\![\hat{n} \in \hat{\omega}]\!]^B] = 1.$$

We also know that $P = C(\omega \times \omega_\alpha, 2)$ is a basis for B. Moreover, it is easy to verify that, for $p \in P$,

$$p \Vdash \hat{n} \in u_\nu \text{ iff } p(n, \nu) = 1;$$
$$p \Vdash \hat{n} \notin u_\nu \text{ iff } p(n, \nu) = 0.$$

We claim that, if $\mu, \nu < \omega_\alpha$ and $\mu \neq \nu$, then $[\![u_\mu = u_\nu]\!]^B = 0$. For suppose not; then there are $\mu, \nu < \omega_\alpha, \mu \neq \nu$ and $p \in P$ such that $p \Vdash u_\mu = u_\nu$. Choose $n \in \omega$ so that $\langle n, \xi \rangle \notin \mathrm{dom}(p)$ for any $\xi < \omega_\alpha$ (possible, since $\mathrm{dom}(p)$ is finite!) and put

$$p' = p \cup \{ \langle \langle n, \mu \rangle, 1 \rangle \} \cup \{ \langle \langle n, \nu \rangle, 0 \rangle \}.$$

Then $p' \Vdash \hat{n} \in u_\mu \wedge \hat{n} \notin u_\nu$, whence $p' \Vdash u_\mu \neq u_\nu$. But since $p' \leq p$ and $p \Vdash u_\mu = u_\nu$, it follows that $p' \Vdash u_\mu = u_\nu$. This contradiction proves the claim.
   Now define $f \in V^{(B)}$ by

$$f = \{ \langle \hat{\nu}, u_\nu \rangle^{(B)} : \nu < \omega_\alpha \} \times \{1\}.$$

One then easily verifies (*cf.* proof of Lemma 1.52) that

$$V^{(B)} \models f \text{ is a map of } \hat{\omega}_\alpha \text{ into } P\hat{\omega}.$$

Moreover, since $[\![u_\mu = u_\nu]\!]^B = 0$ for $\mu \neq \nu$, it quickly follows that $V^{(B)} \models f$ *is one–one*. Since B satisfies the ccc, we have, by Theorem 1.51, $V^{(B)} \models \hat{\omega}_\alpha = \omega_{\hat{\alpha}}$, so that

$$V^{(B)} \models f \text{ is a one-one map of } \omega_{\hat{\alpha}} \text{ into } P\hat{\omega}.$$

Hence $V^{(B)} \models \aleph_{\hat{\alpha}} \leq 2^{\aleph_0}$, and we are done. $\qquad\square$

In the proof of this last theorem we remarked that, for $p \in P = C(\omega \times \omega_\alpha, 2)$, we have $p \Vdash \hat{n} \in u_\nu$ iff $p(n, \nu) = 1$; $p \Vdash \hat{n} \in u_\nu$ iff $p(n, \nu) = 0$. Thus each condition $p \in P$ may be thought of as encoding a finite 'piece of information' about the members of the $\aleph_\alpha$ 'new subsets' u_ν of ω. We may therefore regard $C(\omega \times \omega_\alpha, 2)$ as *the set of conditions for adjoining $\aleph_\alpha$ new subsets of ω using finite pieces of information*, and its completion $\mathrm{RO}(2^{\omega \times \omega_\alpha})$ as an *algebra which adjoins $\aleph_\alpha$ new subsets of ω*.

Corollary 2.13 *If ZF is consistent, so is* $\mathrm{ZFC} + 2^{\aleph_0} = \aleph_2$.

Proof In $\mathrm{ZFC} + \mathrm{GCH}$ we have $\aleph_2^{\aleph_0} = (2^{\aleph_1})^{\aleph_0} = 2^{\aleph_1} = \aleph_2$, so by Theorem 2.12 one can prove in $\mathrm{ZFC} + \mathrm{GCH}$ the existence of a complete Boolean algebra B such that $V^{(B)} \models 2^{\aleph_0} = \aleph_2$. Since $Consis(\mathrm{ZF}) \to Consis(\mathrm{ZFC} + \mathrm{GCH})$, the required result now follows from Theorems 1.33 and 1.19. $\qquad\square$

Similar arguments show that in Corollary 2.13 '$\aleph_2$' can be replaced by '$\aleph_3$', '$\aleph_{\omega+1}$', '$\aleph_{\omega_1}$', etc.

Problems

Throughout, κ denotes an infinite cardinal and B a complete Boolean algebra. In Problems 2.14, 2.15, and 2.16, for any cardinal λ, λ^κ, and $\hat{\lambda}^{\hat{\kappa}}$ are understood to denote, respectively, the set of all maps of κ into λ, and 'the set of all maps of $\hat{\kappa}$ into $\hat{\lambda}$ in $V^{(B)}$'.

2.14 (Infinite distributive laws and $V^{(B)}$) Let λ be a cardinal (finite or infinite). B is said to be (κ, λ)-*distributive* if for any double sequence $\{ b_{\alpha\beta} : \langle \alpha, \beta \rangle \in \kappa \times \lambda \} \subseteq B$ we have

$$\bigwedge_{\alpha < \kappa} \bigvee_{\beta < \lambda} b_{\alpha\beta} = \bigvee_{f \in \lambda^\kappa} \bigwedge_{\alpha < \kappa} b_{\alpha f(\alpha)}.$$

Show that the following conditions are equivalent:

(i) B is (κ, λ)-distributive;

(ii) $V^{(B)} \models \hat{\lambda}^{\hat{\kappa}} = (\lambda^\kappa)\hat{}$.

(Use the fact that

$$\llbracket h \in \hat{\lambda}^{\hat{\kappa}} \rrbracket = \bigwedge_{\alpha < \kappa} \bigvee_{\beta < \lambda} b_{\alpha\beta} \text{ and } \llbracket h \in (\lambda^{\kappa})\,\hat{} \rrbracket = \bigvee_{f \in \lambda^{\kappa}} \bigwedge_{\alpha < \kappa} b_{\alpha f(\alpha)},$$

where $b_{\alpha\beta} = \llbracket h(\hat{\alpha}) = \hat{\beta} \rrbracket$.)

2.15 (Infinite distributive laws and $V^{(B)}$ continued)

(i) B is said to satisfy the *restricted $(\kappa, 2)$-distributive law* if for each double sequence $\{b_{\alpha n} : \langle \alpha, n \rangle \in \kappa \times 2\} \subseteq B$ such that $b_{\alpha 0} = b_{\alpha 1}^{*}$ for all $\alpha < \kappa$, we have $\bigvee_{f \in 2^{\kappa}} \bigwedge_{\alpha < \kappa} b_{\alpha f(\alpha)} = 1$.

Show that the following conditions are equivalent:

(a) B satisfies the restricted $(\kappa, 2)$-distributive law;
(b) B is $(\kappa, 2)$-distributive;
(c) B is (κ, κ)-distributive;
(d) B is $(\kappa, 2^{\kappa})$-distributive;
(e) $V^{(B)} \models \hat{\kappa}^{\hat{\kappa}} = (\kappa^{\kappa})\hat{}\,$;
(f) $V^{(B)} \models P\hat{\kappa} = (P\kappa)\hat{}\,$.

(Show that (a) $\rightarrow$ (f) $\rightarrow$ (e) $\rightarrow$ (d) $\rightarrow$ (c) $\rightarrow$ (b) $\rightarrow$ (a). For (a) $\rightarrow$ (f), use (a) to prove $V^{(B)} \models u \in (P\kappa)\,\hat{}$ for any $u \in B^{\mathrm{dom}(\hat{\kappa})}$. For (f) $\rightarrow$ (e), use the fact that $\kappa^{\kappa} \subseteq P(\kappa \times \kappa)$ *and* $|\kappa| = |\kappa \times \kappa|$. For (e) $\rightarrow$ (d), observe that $(2^{\kappa})^{\kappa} = 2^{\kappa \times \kappa}$ and use Problem 2.14. The remaining implications are trivial.)

(ii) Show that $\mathrm{RO}(2^{\omega})$ is not $(\omega, 2)$-distributive. (Use (i) and Theorem 2.6(i).)

2.16 (Weak distributive laws and $V^{(B)}$) B is said to be *weakly (ω, κ)-distributive* if for each double sequence $\{b_{n\alpha} : \langle n, \alpha \rangle \in \omega \times \kappa\}$ we have

$$\bigwedge_{n \in \omega} \bigvee_{\alpha < \kappa} b_{n\alpha} = \bigvee_{f \in \kappa^{\omega}} \bigwedge_{n \in \omega} \bigvee_{\alpha \leq f(n)} b_{n\alpha}.$$

Clearly, if B is (ω, κ)-distributive, B is weakly (w, κ)-distributive. (But the converse fails: for example, if B is the complete Boolean algebra of Lebesgue measurable subsets of $[0,1]$ modulo the ideal of sets of measure 0, then it can be shown that B is weakly (w, w)-distributive but not (w, w)-distributive.)

We define the *cofinality* $\mathrm{cf}(\kappa)$ of κ to be the least ordinal which is the order type of a cofinal subsct of κ. Thus $\mathrm{cf}\,(\kappa) > w$ iff $\forall f \in \kappa^{\omega} \exists \beta < \kappa \forall n \in \omega [f(n) \leq \beta]$, that is, iff each function from w to κ is bounded by some ordinal $< \kappa$.

(i) Show that, if $\mathrm{cf}\,(\kappa) > w$, then

$$\bigvee_{f \in \kappa^{\omega}} \bigwedge_{n \in \omega} \bigvee_{\alpha \leq f(n)} b_{n\alpha} = \bigvee_{\beta < \kappa} \bigwedge_{n \in \omega} \bigvee_{\alpha \leq \beta} b_{n\alpha}.$$

(ii) Show that, if B satisfies ccc and cf $(\kappa) > \omega$, then B is weakly (ω, κ)-distributive. (Let $\{b_{n\alpha} : \langle n, \alpha \rangle \in \omega \times \kappa\} \subseteq B$. Using Problem 1.53(iv), replace each $\{b_{n\alpha} : \alpha < \kappa\}$ by a countable subset C_n having the same supremum.)

(iii) Suppose that cf $(\kappa) > \omega$. Show that B is weakly (ω, κ)-distributive iff $V^{(B)} \models \mathrm{cf}\hat{\kappa} > \hat{\omega}$. (Note that $[\![f \in \hat{\kappa}^{\hat{\omega}}]\!] = \bigwedge_{n \in \omega} \bigvee_{\alpha < \kappa} b_{n\alpha}$ and that $[\![\exists \beta < \hat{\kappa} \forall n \in \hat{\omega}[f(n) \le \beta]]\!] = \bigvee_{\beta < \kappa} \bigwedge_{n \in \omega} \bigvee_{\alpha \le \beta} b_{n\alpha}$, where $b_{n\alpha} = [\![f(\hat{n}) = \hat{\alpha}]\!]$.)

(iv) Show that B is weakly (ω, ω)-distributive iff $V^{(B)} \models \forall g[g \in \hat{\omega}^{\hat{\omega}} \to \exists f \in (\omega^\omega)^\frown \forall n \in \hat{\omega}[g(n) \le f(n)]$, in other words; iff in $V^{(B)}$ the *standard* numerical functions are cofinal in the class of *all* numerical functions. (Argue as in (iii).)

2.17 (κ-closure and $V^{(B)}$) Let P be a basis for B. P is said to be *κ-closed* if for each ordinal $\alpha < \kappa$ and each descending α-sequence $p_0 \ge p_1 \ge \cdots \ge p_\beta \ge \cdots (\beta < \alpha)$ in P there is $p \in P$ such that $p \le p_\beta$ for all $\beta < \alpha$. Consider the following conditions:

(i) B has a dense subset P which is κ-closed;

(ii) for any $\alpha < \kappa$ and any $x \in V$,

$$V^{(B)} \models \hat{x}^{\hat{\alpha}} = (x^\alpha)\,\widehat{}\,;$$

(iii) $V^{(B)} \models \mathrm{Card}(\hat{\alpha})$ for any cardinal $\alpha \le \kappa$;

(iv) $V^{(B)} \models P\hat{\alpha} = (P\alpha)\widehat{}$ for any $\alpha \le \kappa$.

Show that (i) $\to$ (ii) $\to$ (iii), and (ii) $\to$ (iv). Hence (i) implies that B is (α, λ)-distributive for any $\alpha < \kappa$ and any λ. (For (i) $\to$ (ii), let $p \in P$ be such that $p \Vdash f \in \hat{x}^{\hat{\alpha}}$. Using (i), find a descending sequence $\{p_\beta : \beta < \alpha\} \subseteq P$ and a set $\{y_\beta : \beta < \alpha\} \subseteq x$ such that $p_\beta \Vdash f(\hat{\beta}) = \hat{y}_\beta$. If $q \in P$ satisfies $q \le p_\beta$ for all $\beta < \alpha$, and $g = \{\langle \beta, y_\beta \rangle : \beta < \alpha\}$, show that $q \Vdash f = \hat{g}$.)

2.18 (An important set of conditions) Let x and y be nonempty sets, where $|y| \ge 2$. We put $C_\kappa(x, y)$ for the set—partially ordered by $\supseteq$—of all maps with domain a subset of x of cardinality $< \kappa$ and range a subset of y. Put $B_\kappa(x, y)$ for the regular open algebra of the space y^x with the topology whose basic open sets are of the form

$$N(p) = \{f \in y^x : p \subseteq f\}$$

for $p \in C_\kappa(x, y)$. Observe that $C_\omega(x, y) = C(x, y)$.

(i) Show that $C_\kappa(x, y)$ is refined, and that $\langle B_\kappa(x, y), N \rangle$ is a Boolean completion of $C_\kappa(x, y)$. Thus $C_\kappa(x, y)$ is an basis for $B_\kappa(x, y)$.

(ii) Show that, if κ is regular, $C_\kappa(x, y)$ is κ-closed.

(iii) Assume GCH, κ is regular and $|y| \leq \kappa$. Show that, if I is any set of pairwise incompatible elements of $C_\kappa(x, y)$, then $|I| \leq \kappa$. (Fix a well-ordering of I. Define a sequence $\{x_\alpha : \alpha < \kappa^+\}$ of subsets of x by: $x_0 = \emptyset; x_\alpha = \bigcup_{\beta < \alpha} x_\beta$ for limit α, $x_{\alpha+1} = x_\alpha \cup \bigcup \{\mathrm{dom}(q) :$ for some $p \in C_\kappa(x_\alpha, y), q$ is the least element of I such that $q|x_\alpha = p\}$. Show that $|x_\alpha| \leq \kappa$ for $\alpha < \kappa^+$, and hence that $|C_\kappa(x_\kappa, y)| \leq \kappa$. Now prove that $I \subseteq C_\kappa(x_\kappa, y)$: for any $p \in I$, Show that there is $\alpha < \kappa$ such that $\mathrm{dom}(p) \cap x_\alpha = \mathrm{dom}(p) \cap x_{\alpha+1}$; choose $q \in I$ such that $p|x_\alpha = q|x_\alpha$ and $\mathrm{dom}(q) \subseteq x_{\alpha+1}$; show that p and q coincide on $\mathrm{dom}(p) \cap \mathrm{dom}(q)$; deduce that $p = q$ and $p \in C_\kappa(x_\kappa, y)$.)

(iv) Assume GCH, κ is regular and $|y| \leq \kappa$. Show that $B_\kappa(x, y)$ satisfies $\kappa^+ - cc$. (Use (iii).)

(v) Assume GCH, κ and $|x|$ are regular and $\kappa < |x|$. Show that $|B_\kappa(x, 2)| = |x|$. (Using (iii), argue as in Lemma 2.10.)

2.19 (Consistency of $2^{\aleph_0} = \aleph_2 + \forall \kappa \geq \aleph_1 [2^\kappa = \kappa^+]$ with ZFC)

(i) Show that, if $|B| = \lambda$, then $V^{(B)} \models |P\hat\kappa| \leq |(\lambda^\kappa)\hat{\ }|$. (Use Lemma 1.52.)

(ii)[1] Assume GCH, and let $|B| = \lambda \geq \aleph_0$. Show that $V^{(B)} \models \forall_\alpha \geq \hat\lambda [\mathrm{Card}(\alpha) \to 2^\alpha = \alpha^+]$. Use Problem 1.53 and (i).)

(iii) Assume GCH, and let $B = \mathrm{RO}(2^{\omega \times \omega_2})$. Show that $V^{(B)} \models 2^{\aleph_1} = \aleph_2$ (Use (i)) and deduce from this, Theorem 2.12 and (ii) that if ZF is consistent, so is ZFC $+ 2^{\aleph_0} = \aleph_2 + \forall \kappa \geq \aleph_1 [2^\kappa = \kappa^+]$.

2.20 (A further relative consistency result) Assume GCH. Let κ, λ be regular cardinals such that $\kappa < \lambda$. Put $B = B_\kappa(\kappa \times \lambda, 2)$ (Problem 2.18).

(i) Show that $V^{(B)} \models \mathrm{Card}(\hat\alpha)$ for any cardinal α. (For $\alpha \leq \kappa$, use Problems 2.18(ii) and 2.17(iii). For $\alpha \geq \kappa^+$, use Problems 2.18(iv) and 1.53).

(ii) Show that $V^{(B)} \models P\hat\alpha = (P\alpha)\hat{\ }$ for any cardinal $\alpha < \kappa$. (Use Problems 2.18(ii) and 2.17(iv).)

(iii) Show that

$$V^{(B)} \models \forall_\alpha < \hat\kappa [\mathrm{Card}(\alpha) \wedge \aleph_0 \leq \alpha \to 2^\alpha = \alpha^+].$$

(Use (i), (ii), and GCH.)

(iv) Show that

$$V^{(B)} \models \forall_\alpha \geq \hat\lambda [\mathrm{Card}(\alpha) \to 2^\alpha = \alpha^+].$$

(Use Problems 2.18(v), 2.19(i), and (i).)

---

[1] This result shows that, as long as B is a *set* (i.e. has a cardinality) we cannot *provably* violate the GCH in $V^{(B)}$ at arbitrarily high cardinals. It turns out, however, that this can be achieved when B is a suitably chosen (proper) *class*. The details are, unfortunately, too lengthy to be included here. See Easton (1970), Takeuti and Zaring (1973), or Shoenfield (1971).

(v) Show that

$$V^{(B)} \models \forall \alpha[\mathrm{Card}(\alpha) \wedge \hat{\kappa} \leq \alpha < \hat{\lambda} \to 2^{\alpha} = \hat{\lambda}].$$

(Use Problems 2.18(iv) and 2.19(i) to show that $V^{(B)} \models 2^{\hat{\alpha}} \leq \hat{\lambda}$ for $\kappa \leq \alpha < \lambda$ and argue as in the proof of Theorem 2.12 to get $V^{(B)} \models 2^{\hat{\kappa}} \geq \hat{\lambda}$.)

(vi) Deduce that, if ZF is consistent, so is

$$\mathrm{ZFC} + 2^{\aleph_0} = \aleph_1 + \forall \kappa[\aleph_1 \leq \kappa \leq \aleph_\omega \to 2^{\kappa} = \aleph_{\omega+1}]$$
$$+ \forall \kappa[\aleph_{\omega+1} \leq \kappa \to 2^{\kappa} = \kappa^+].$$

2.21 (Consistency of $\mathrm{GCH} + P\omega \subseteq L + P\omega_1 \nsubseteq L$ **with ZFC)** Assume GCH, let κ be regular, and put $B = B_\kappa(\kappa, 2)$.

(i) Show that $V^{(B)} \models \mathrm{Card}(\hat{\alpha})$ for any cardinal α. (Like Problem 2.20(i).)

(ii) Show that $V^{(B)} \models P\hat{\alpha} = (P\alpha)^{\hat{}}$ for any cardinal $\alpha < \kappa$. (Like Problem 2.20(ii).)

(iii) Show that $V^{(B)} \models P\hat{\kappa} \neq (P\kappa)^{\hat{}}$. (Like Theorem 2.6(i).)

(iv) Show that $V^{(B)} \models \mathrm{GCH}$. (To show $V^{(B)} \models 2^{\hat{\alpha}} = \hat{\alpha}^+$ for $\alpha < \kappa$, argue as in Problem 2.20(iii). (To show $V^{(B)} \models 2^{\hat{\alpha}} = \hat{\alpha}^+$ for $\alpha \geq \kappa$, argue as in Problem 2.20(iv).)

(v) Assume $V = L$. Show that

$$V^{(B)} \models \forall \alpha < \hat{\kappa}[P\alpha \subseteq L] \wedge P\hat{\kappa} \nsubseteq L.$$

(Using (i) and (ii), argue as in Theorem 2.6.)

(vi) Deduce that, if ZF is consistent, so is $\mathrm{ZFC} + \mathrm{GCH} + P\omega \subseteq L + P\omega_1 \nsubseteq L$.

3

GROUP ACTIONS ON $V^{(B)}$ AND THE INDEPENDENCE OF THE AXIOM OF CHOICE

Group actions on $V^{(B)}$

Let G be a group, and X a class. An *action* of G on X is a map $\langle g, x \rangle \mapsto g \cdot x : G \times X \to X$ satisfying

$$1 \cdot x = x \quad \text{and} \quad (gh) \cdot x = g \cdot (h \cdot x)$$

for all $x \in X$, g, $h \in G$, where 1 is the identity element of G. (When confusion is unlikely, we write gx for $g \cdot x$.) Under these conditions we say that G *acts* on X. For each $g \in G$, the map $\pi_g : X \to X$ defined by $\pi_g(x) = g \cdot x$ is a permutation of X, and the correspondence $g \mapsto \pi_g$ defines a homomorphism of G into the group of permutations of X.

If B is a Boolean algebra, by an action of a group G on B we shall always mean an *action of G by automorphisms*, that is, one in which each π_g as defined above is not merely a permutation but actually an *automorphism* of B. In particular, the automorphism group $\mathrm{Aut}(B)$ of B acts on B in the natural way via:

$$\pi \cdot b = \pi(b)$$

for $\pi \in \mathrm{Aut}(B)$, $b \in B$.

We can extend the notion of group actions to *Boolean-valued structures* as follows. Let B be a complete Boolean algebra, and let $\mathcal{S} = \langle S, [\![\cdot = \cdot]\!]_S, [\![\cdot \in \cdot]\!]_S \rangle$ be a B-valued structure. An *action* of a group G on $\mathcal{S}$ is a pair of actions of G on B and the class S satisfying

$$\begin{aligned}
[\![gu = gv]\!]_S &= g \cdot [\![u = v]\!]_S \\
[\![gu \in gv]\!]_S &= g \cdot [\![u \in v]\!]_S.
\end{aligned} \tag{3.1}$$

It is easily shown by induction on the complexity of formulas that for any formula $\phi(v_1, \ldots, v_n)$ of $\mathcal{L}$, any $x_1, \ldots, x_n \in S$ and any $g \in G$,

$$[\![\phi(gx_1, \ldots, gx_n)]\!]_S = g \cdot [\![\phi(x_1, \ldots, x_n)]\!]_S. \tag{3.2}$$

We now show that any action of a group G on a complete Boolean algebra B extends naturally to an action of G on the B-valued structure $V^{(B)}$.

Theorem 3.3 *Let G be a group acting on the complete Boolean algebra B. Define the map $\langle g, u \rangle \mapsto gu : G \times V^{(B)} \to V^{(B)}$ by recursion on the well-founded relation $y \in \mathrm{dom}(x)$ via*

$$gu = \{\langle gx, g \cdot u(x) \rangle : x \in \mathrm{dom}(u)\}. \tag{3.4}$$

Then this defines an action of G on $V^{(B)}$ such that:

(i) *for any $u \in V^{(B)}$, $g \in G$, we have $\mathrm{dom}(gu) = \{gx : x \in \mathrm{dom}(u)\}$ and, for any $x \in \mathrm{dom}(u)$, $(gu)(gx) = g \cdot u(x)$;*

(ii) *$g\hat{v} = \hat{v}$ for any $v \in V$.*

Proof To show that (3.4) defines an action of G on $V^{(B)}$, we first prove that for any $g \in G$ the map $x \mapsto gx$ for $x \in V^{(B)}$ is one–one and sends $V^{(B)}$ into itself. For this it suffices to show by induction on α that

(1) *the restriction to $V_\alpha^{(B)}$ of $x \mapsto gx$ is one–one from $V_\alpha^{(B)}$ to $V^{(B)}$.*

Assume that (1) holds for all $\beta < \alpha$. If $u \in V_\alpha^{(B)}$ then $\mathrm{dom}(u) \subseteq V_\beta^{(B)}$ for some $\beta < \alpha$, so that the restriction of $x \mapsto gx$ to $\mathrm{dom}(u)$ is a one–one map of $\mathrm{dom}(u)$ into $V^{(B)}$. It follows immediately that gu as defined by (3.4) is a map of $\{gx : x \in \mathrm{dom}(u)\} \subseteq V^{(B)}$ into B, so that $gu \in V^{(B)}$. Thus the restriction to $V_\alpha^{(B)}$ of $x \mapsto gx$ carries $V_\alpha^{(B)}$ into $V^{(B)}$. To show that it is one–one, suppose that $u, v \in V_\alpha^{(B)}$ and $gu = gv$. Then, by (3.4),

(2) $\{\langle gx, g \cdot u(x) \rangle : x \in \mathrm{dom}(u)\} = \{\langle gy, g \cdot v(y) \rangle : y \in \mathrm{dom}(v)\}.$

But there is $\beta < \alpha$ such that $V_\beta^{(B)}$ includes both $\mathrm{dom}(u)$ and $\mathrm{dom}(v)$, so that $x \mapsto gx$ is one–one on both these sets. It now follows from (2) that

$$\{\langle x, u(x) \rangle : x \in \mathrm{dom}(u)\} = \{\langle y, v(y) \rangle : y \in \mathrm{dom}(v)\}$$

that is, $u = v$. Hence the restriction to $V_\alpha^{(B)}$ of $x \mapsto g \cdot x$ is one–one and (1) is proved.

Part (i) now follows immediately from (3.4).

To establish that $\langle g, u \rangle \mapsto gu$ is an action of G on $V^{(B)}$, we first use the induction principle for $V^{(B)}$ to show that $(gh)u = g(hu)$ for any $g, h \in G$, $u \in V^{(B)}$. Assuming accordingly that $(gh)x = g(hx)$ for all $x \in \mathrm{dom}(u)$, we compute

$$g(hu) = \{\langle gy, g(hu)(y) \rangle : y \in \mathrm{dom}(hu)\}$$
$$= \{\langle g(hx), g(hu)(hx) \rangle : x \in \mathrm{dom}(u)\}$$
$$= \{\langle g(hx), g(h \cdot u(x)) \rangle : x \in \mathrm{dom}(u)\}$$
$$= \{\langle (gh)x, (gh) \cdot u(x) \rangle : x \in \mathrm{dom}(u)\}$$
$$= (gh)u,$$

which proves the assertion. Similarly, one shows that $1u = u$ for all $u \in V^{(B)}$. The facts that

$$g \cdot [\![u \in v]\!] = [\![gu \in gv]\!]$$

and

$$g \cdot [\![u = v]\!] = [\![gu = gv]\!]$$

are proved simultaneously using the induction principle for $V^{(B)}$. We omit the straightforward details.

Finally (ii) is proved by a simple induction on the well-founded relation $y \in x$.
$\square$

Recall that the automorphism group $\mathrm{Aut}(B)$ of B acts on B; hence it also acts on $V^{(B)}$. An element x of B or of $V^{(B)}$ is said to be *invariant* if $\pi x = x$ for every $\pi \in \mathrm{Aut}(B)$. B is said to be *homogeneous* if 0 and 1 are its only invariant elements.

Problem 3.5 (Another characterization of homogeneity) Show that B is homogeneous iff for each $x \neq 0$, $y \neq 0$ in B there is an automorphism π of B such that $x \wedge \pi y \neq 0$. (Consider $\bigvee \{\pi y : \pi$ an automorphism of $B\}$.)

Next we show that homogeneity of B confers certain desirable properties on $V^{(B)}$.

Lemma 3.6 *Suppose that B is homogeneous. Then, for any formula $\phi(v_1, \ldots, v_n)$ and any $x_1, \ldots, x_n \in V$, either $[\![\phi(\hat{x}_1, \ldots, \hat{x}_n)]\!]^B = 0$ or else $[\![\phi(\hat{x}_1, \ldots, \hat{x}_n)]\!]^B = 1$. In particular, for any sentence σ, either $[\![\sigma]\!]^B = 0$ or $[\![\sigma]\!]^B = 1$.*

Proof By (3.2) and Theorem 3.3(ii) $[\![\phi(\hat{x}_1, \ldots, \hat{x}_n)]\!]^B$ is an invariant element of B. The result now follows from the homogeneity of B.
$\square$

To conclude this section, we establish the existence of a large class of homogeneous algebras.

Lemma 3.7 *For any set I, $\mathrm{RO}(2^I)$ is homogeneous.*

Proof For each $i \in I$ define $\pi_i : 2^I \to 2^I$ by $\pi_i f = f^i$ for $f \in 2^I$, where $f^i \in 2^I$ is defined by

$$f^i(j) = f(j) \text{ for } j \neq i$$
$$f^i(i) = 1 - f(i).$$

It is easy to check that π_i is a homeomorphism of 2^I onto itself, and so induces an automorphism π_i' of $\mathrm{RO}(2^I) = B$ defined by

$$\pi_i'(U) = \pi_i^{-1}[U]$$

for $U \in B$.

Suppose now that U is an invariant element of B. Then if $U \neq \emptyset$ there is a basic open set

$$\bigcap_{k=1}^{n} \{f \in 2^I : f(i_k) = a_k\} \subseteq U, \tag{1}$$

where $\{i_1, \ldots, i_n\} \subseteq I$ and $\{a_1, \ldots, a_n\} \subseteq 2$. Applying π_{i_n}' and using the invariance of U, it follows that

$$\bigcap_{k=1}^{n-1} \{f \in 2^I : f(i_k) = a_k\} \cap \{f \in 2^I : f(i_n) = 1 - a_n\} \subseteq U.$$

This, together with (1) implies

$$\bigcap_{k=1}^{n-1} \{f \in 2^I : f(i_k) = a_k\} \subseteq U.$$

Continuing in this way we get $2^I \subseteq U$, so that $U = 1$ in B. The homogeneity of B follows. $\qquad\square$

The independence of the existence of definable well-orderings of $P\omega$

We now apply the results of the previous section to show that, if ZF is consistent, so is ZFC + GCH + 'there is no definable well-ordering of $P\omega$'. Thus, although in ZFC one can prove the *existence* of a well-ordering of $P\omega$, even in the presence of GCH it is consistent to assume that no such well-ordering can be *explicitly defined*.

For each formula $\phi(x, y)$ let WO_ϕ be the sentence 'ϕ defines a well-ordering of $P\omega$'. Then we have

Theorem 3.8 *Let $B = \mathrm{RO}(2^\omega)$. Then for any formula $\phi(x, y)$ we have*

$$V^{(B)} \models \neg\mathrm{WO}_\phi.$$

Proof Before launching into formalities we give an outline of the proof. By Theorem 2.6 $P\hat{\omega} - (P\omega)\hat{\ }$ is nonempty in $V^{(B)}$. If $P\hat{\omega}$ had a definable well-ordering in $V^{(B)}$, $P\hat{\omega} - (P\omega)\hat{\ }$ would have a definable least element u. But then u would be invariant, and one can use the homogeneity of B to show that $u \in (P\omega)\hat{\ }$ in $V^{(B)}$, a contradiction.

Now for the formal details. Put $c = [\![\mathrm{WO}_\phi]\!]^B$ and write S for $P\hat{\omega} - (P\omega)\hat{\ }$ in $V^{(B)}$. Then, by Lemmas 3.7 and 3.6, either $c = 0$ or $c = 1$; we have to prove the former. Suppose, for contradiction's sake, that $c = 1$. By Theorem 2.6 we have $V^{(B)} \models S \neq \emptyset$, so it follows that

(1) $V^{(B)} \models \exists! x \in S \forall y \in S \phi(x, y).$

Hence, by the Maximum Principle, there is $u \in V^{(B)}$ such that

(2) $V^{(B)} \models u \in S \wedge \forall y \in S \phi(u, y).$

We have $V^{(B)} \models u \subseteq \hat{\omega}$ and, by (1) and (2), for all $n \in \omega$,

$$V^{(B)} \models [\hat{n} \in u \leftrightarrow \exists x \in S[\hat{n} \in x \wedge \forall y \in S \phi(x, y)]].$$

Hence

$$[\![\hat{n} \in u]\!]^B = [\![\exists x \in S[\hat{n} \in x \wedge \forall y \in S \phi(x, y)]]\!]^B.$$

Now the r.h.s. of this equation is evidently invariant, so by Lemmas 3.6 and 3.7 the l.h.s. is either 0 or 1. Put

$$v = \{n \in \omega : [\![\hat{n} \in u]\!]^B = 1\}.$$

Then $[\![\hat{n} \in \hat{v}]\!]^B = [\![\hat{n} \in u]\!]^B$, so that

$$[\![\forall x \in \hat{\omega}[x \in u \leftrightarrow x \in \hat{v}]]\!]^B = 1,$$

Whence $V^{(B)} \models u = \hat{v}$, and thus $V^{(B)} \models u \in (P\omega)\hat{\ }$. But this contradicts the fact—immediate from (2)—that $V^{(B)} \models u \notin (P\omega)\hat{\ }$. Thus $c = 0$ and we are through. $\qquad\square$

Corollary 3.9 *If* ZF *is consistent, so is*

$$\mathrm{ZFC} + \mathrm{GCH} + \{\neg\mathrm{WO}_\phi : \phi(x, y) \text{ a formula}\}.$$

Proof If $B = \mathrm{RO}(2^\omega)$ then $|B| = 2^{\aleph_0}$ and so by 2.8 in $\mathrm{ZFC} + \mathrm{GCH}$ we can prove that $V^{(B)} \models \mathrm{GCH}$. The required result now follows easily from this, Theorems 3.8, 1.33, and 1.19. $\qquad\square$

Problems

3.10 (The Boolean-valued subset defined by a formula) Let $\psi(x)$ be any B-formula and let $u \in V^{(B)}$. Recall that in the proof of Lemma 1.35 we showed that the object $v \in V^{(B)}$ defined by $\mathrm{dom}(v) = \mathrm{dom}(u)$ and $v(x) = u(x) \wedge [\![\psi(x)]\!]$

for $x \in \mathrm{dom}(v)$ satisfies $V^{(B)} \models \forall x[x \in v \leftrightarrow x \in u \wedge \psi(x)]$. Write $v = \{x \in u : \psi(x)\}^{(B)}$. Now suppose that B is homogeneous, $\phi(x, v_1, \ldots, v_n)$ is any formula and $a, a_1, \ldots, a_n \in V$. Show that

$$V^{(B)} \models \{x \in \hat{a} : \phi(x, \hat{a}_1, \ldots, \hat{a}_n)\}^{(B)} \in (Pa)\hat{\;}.$$

3.11 (Ordinal definable sets in $V^{(B)}$) We recall (*cf.* Drake 1974, ch. 5) that a set u is said to be *ordinal definable*—written $\mathrm{OD}(u)$—if for some formula ϕ and ordinals $\alpha_1, \ldots, \alpha_n, u$ is the unique set such that $\phi(u, \alpha_1, \ldots, \alpha_n)$ holds. The set u is *hereditarily ordinal definable*—written $\mathrm{HOD}(u)$—if all the members of the transitive closure of $\{u\}$ are ordinal definable. We put $\mathrm{OD} = \{u : \mathrm{OD}(u)\}$ and $\mathrm{HOD} = \{u : \mathrm{HOD}(u)\}$. Recall the following facts:

(a) OD has a definable well-ordering and, for any set $u, u \subseteq \mathrm{OD}$ iff u has a definable well-ordering;

(b) $L \subseteq \mathrm{HOD} \subseteq \mathrm{OD}$.

   (i) Show that $L = \mathrm{HOD}$ iff for all u, $(u \subseteq L \wedge u \in \mathrm{OD}) \rightarrow u \in L$. (Use $\in$-induction.)

   (ii) Suppose that B is homogeneous. Show that $V = L \rightarrow V^{(B)} \models L = \mathrm{HOD}$. (Use (i) and Problem 3.10.)

   (iii) Show that, if ZF is consistent, so is $\mathrm{ZFC} + \mathrm{GCH} + L = \mathrm{HOD} + \mathrm{HOD} \neq V$. (Use (ii) and Corollary 3.9.)

3.12 (Complete homomorphisms) Let B and B' be complete Boolean algebras. Recall that a homomorphism $h : B \rightarrow B'$ is said to be *complete* if $h(\bigvee X) = \bigvee h[X]$ for any $X \subseteq B$.

   (i) Let h be a complete *monomorphism* of B into B'. Define the map $\bar{h}$ on $V^{(B)}$ by recursion: for all $u \in V^{(B)}$

$$\bar{h}u = \{\langle \bar{h}x, h(u(x))\rangle : x \in \mathrm{dom}(u)\}.$$

Show that $\bar{h}$ is an injection of $V^{(B)}$ into $V^{(B')}$ such that, for any $u, v \in V^{(B)}$, $h[\![u \in v]\!]^B = [\![\bar{h}u \in \bar{h}v]\!]^{B'}$, $h[\![u = v]\!]^B = [\![\bar{h}u = \bar{h}v]\!]^{B'}$ and, for $x \in V$, $\bar{h}\hat{x} = \hat{x}$. (Argue inductively as in the proof of Theorem 3.3.)

Throughout the remainder of this problem take h to be a complete homomorphism of B into B'.

   (ii) Define the map $\tilde{h}$ on $V^{(B)}$ by recursion: for $u \in V^{(B)}$

$$\tilde{h}u = \left\{ \left\langle \tilde{h}x, \bigvee_{B'} \{h(u(y)) : y \in \mathrm{dom}(u) \wedge \tilde{h}x = \tilde{h}y\} \right\rangle : x \in \mathrm{dom}(u) \right\}.$$

Show that (a) $\tilde{h} : V^{(B)} \to V^{(B')}$, (b) $\tilde{h}$ is onto if h is, (c) $\tilde{h} = \bar{h}$ if h is a monomorphism, (d) $\tilde{h}\hat{x} = \hat{x}$ for $x \in V$, (e) $h[\![u \in v]\!]^B = [\![\tilde{h}u \in \tilde{h}v]\!]^{B'}$, $h[\![u = v]\!]^B = [\![\tilde{h}u = \tilde{h}v]\!]^{B'}$ for $u, v \in V^{(B)}$. (Argue inductively.)

(iii) If h' is a complete homomorphism of B' into a complete Boolean algebra B'', show that $(h' \circ h)\tilde{} = \tilde{h}' \circ \tilde{h}$. (Argue inductively.)

(iv) Show that for any Σ_1-formula $\phi(v_1, \ldots, v_n)$ and any $x_1, \ldots, x_n \in V^{(B)}$

$$h[\![\phi(x_1, \ldots, x_n)]\!]^B \leq [\![\phi(\tilde{h}x_1, \ldots, \tilde{h}x_n)]\!]^{B'}.$$

(v) Show that, if h is *onto*, then for any formula $\phi(v_1, \ldots, v_n)$ and any $x_1, \ldots, x_n \in V^{(B)}$,

$$h[\![\phi(x_1, \ldots, x_n)]\!]^{(B)} = [\![\phi(\tilde{h}x_1, \ldots, \tilde{h}x_n)]\!]^{B'}.$$

3.13 (Ultrapowers as Boolean extensions)

(i) Let I be a set and for each $i \in I$ define $\pi_i : PI \to 2$ by $\pi_i(X) = 1$ if $i \in X$, $\pi_i(X) = 0$ if $i \notin X$; then π_i is a complete homomorphism of the complete Boolean algebra PI onto 2. Let $\phi(v_1, \ldots, v_n)$ be a formula and let $x_1, \ldots, x_n \in V^{(PI)}$. Show that

$$[\![\phi(x_1, \ldots, x_n)]\!]^{PI} = \{i \in I : [\![\phi(\tilde{\pi}_i x_1, \ldots, \tilde{\pi}_i x_n)]\!]^2 = 1\}.$$

(Use Problem 3.12(v).)

(ii) Let B be a complete Boolean algebra and let U be an *ultrafilter* in B. Define the relation $\sim_U$ on $V^{(B)}$ by $x \sim_U y \leftrightarrow [\![x = y]\!] \in U$; then $\sim_U$ is an equivalence relation on $V^{(B)}$. For each $x \in V^{(B)}$ let $x^U = \{y \in V^{(B)} : x \sim_U y\}$ be the $\sim_U$-equivalence class of x. Define the relation $\in_U$ on the class[1] $\{x^U : x \in V^{(B)}\}$ by $x^U \in_U y^U \leftrightarrow [\![x \in y]\!] \in U$. Let $V^{(B)}/U$ be the structure $\langle \{x^U : x \in V^{(B)}\}, \in_U \rangle$: this is called the *quotient* of $V^{(B)}$ by U. (For a fuller treatment of quotients, see Chapter 4.) Let $\phi(v_1, \ldots, v_n)$ be a formula and let $x_1, \ldots, x_n \in V^{(B)}$. Show that

$$V^{(B)}/U \models \phi[x_1^U, \ldots, x_n^U] \text{ iff } [\![\phi(x_1, \ldots, x_n)]\!] \in U.$$

(Induction on the complexity of ϕ, using the Maximum Principle to handle the existential case.)

(iii) Let I be a set. Define $g : V^{(2)} \to V$ by putting, for each $u \in V^{(2)}$, $g(u) = $ *the unique* $x \in V$ such that $V^{(PI)} \models u = \hat{x}$ (Theorem 1.23(iv)). Define $h : V^{(PI)} \to V^I$, the set of all maps with domain I, by

---

[1] Strictly speaking, each x^U is itself a (proper) class, so $\{x^U : x \in V^{(B)}\}$ is not defined. However, this annoyance can be overcome by Scott's well-known trick of replacing each x^U by the set of its members of minimum rank.

$h(x) = \{<i, g(\tilde{\pi}_i x)> : i \in I\}$ for $x \in V^{(PI)}$. Show that h is onto. (Start with $f \in V^I$; observe that $\{f^{-1}(f(i)) : i \in I\}$ is a partition of unity in PI. Hence by Problem 1.26(i) there is $x \in V^{(PI)}$ such that $[\![x = f(i)^\frown]\!]^{PI} = f^{-1}(f(i))$ for $i \in I$. Show, using Problem 3.12 and (i), that $h(x) = f$.)

(iv) Let I be a set, let U be an ultrafilter in PI, and let V^I/U be the usual ultrapower of V by U. Define $j : V^{(PI)}/U \to V^I/U$ by $j(x^U) = h(x)/U$, the canonical image of $h(x)$ in V^I/U (and h is defined as in (iii)). Show that j is an isomorphism of $V^{(PI)}/U$ onto V^I/U. (Use (i), (ii), and (iii).) *Thus, each ultrapower of V can be obtained as a quotient of a suitable Boolean extension of V.*

The independence of the axiom of choice

We turn next to the problem of establishing the relative consistency of $\neg$AC with ZF. Now it is clear that we cannot do this by trying to falsify AC in some $V^{(B)}$ (as, for example, we did with CH), because we know that in ZFC one can *prove* that AC is true in $V^{(B)}$. It turns out, however, that, if B is acted on by a suitable *group*, then we can falsify AC in certain *submodels* of $V^{(B)}$. We first give a heuristic sketch of the argument.

Let G be the group of all permutations of ω and for each $n \in \omega$ let

$$G_n = \{g \in G : gn = n\}.$$

We choose a certain complete Boolean algebra B and construct a certain subclass $V^{(\Gamma)}$ of $V^{(B)}$ such that

(i) $V^{(\Gamma)}$ is a B-valued model of ZF such that $\hat{x} \in V^{(\Gamma)}$ for all $x \in V$;

(ii) G acts on $V^{(\Gamma)}$;

(iii) for each $x \in V^{(\Gamma)}$, there is a finite subset $J \subseteq \omega$ (called a *support* of x) such that $gx = x$ for every $g \in \bigcap_{n \in J} G_n$;

(iv) there is an infinite 'set of distinct reals' $\{u_n : n \in \omega\} = s$ in $V^{(\Gamma)}$ such that $gu_n = u_{gn}$ for all $g \in G$.

Then, in $V^{(\Gamma)}$, s is infinite but not Dedekind infinite, so *a fortiori* the axiom of choice fails in $V^{(\Gamma)}$. For suppose f is any map in $V^{(\Gamma)}$ of $\hat{\omega}$ into s. Then, by (iii), f has a finite support J. If f were one–one, then there would be $n \notin J$ such that $u_n \in \mathrm{ran}(f)$. Choose $n' \notin \{n\} \cup J$ and let $g \in G$ be the permutation of ω, which interchanges n and n' but leaves everything else undisturbed. If $u_n = f(\hat{m})$, then $u_{n'} = u_{gn} = gu_n = g(f(\hat{m})) = (gf)(g\hat{m}) = f(\hat{m}) = u_n$, contradicting $u_n \neq u_{n'}$. Thus there is no one–one map of $\hat{\omega}$ into s, so that s is not Dedekind infinite.

Condition (iii) implies that the members of $V^{(\Gamma)}$ have the following property: for $x \in V^{(B)}$ let $\mathrm{stab}(x) = \{g \in G : gx = x\}$; then $\mathrm{stab}(x) \in \Gamma$ for every $x \in V^{(\Gamma)}$, where Γ is the filter of subgroups generated by the G_n, that is, $\Gamma = \{H : H$ *a subgroup of* G *and for some finite* $J \subseteq \omega, \bigcap_{n \in J} G_n \subseteq H\}$. This leads us to consider an (arbitrary) *filter of subgroups* of an (arbitrary) group G.

Finally, since we want G to act on $V^{(\Gamma)}$, we must have $x \in V^{(\Gamma)} \to gx \in V^{(\Gamma)} \to$ stab$(gx) \in \Gamma$. But it is easy to verify that stab$(gx) = g\,\text{stab}(x)g^{-1}$, so we shall want Γ to satisfy $H \in \Gamma \to gHg^{-1} \in \Gamma$ for $g \in G$. Under these conditions Γ is said to be *normal*.

We now turn to the formal development. Let G be a group acting on the complete Boolean algebra B, and let Γ be a *filter of subgroups* of G. That is, Γ is a nonempty set of subgroups of G such that

$$H, K \in \Gamma \to H \cap K \in \Gamma$$

and

$$H \in \Gamma \text{ and } H \subseteq K, \ K \text{ a subgroup of } G \to K \in \Gamma.$$

Γ is called a *normal* filter if

$$g \in G \text{ and } H \in \Gamma \to gHg^{-1} \in \Gamma.$$

By Theorem 3.3, G acts on $V^{(B)}$; for each $x \in V^{(B)}$ we define the *stabilizer* of x by

$$\text{stab}(x) = \{g \in G : gx = x\}.$$

It is easy to verify that stab(x) is subgroup of G. We define (by analogy with (1.4)) the sets $V_\alpha^{(\Gamma)}$ recursively as follows:

$$V_\alpha^{(\Gamma)} = \{x : \text{Fun}(x) \wedge \text{ran}(x) \subseteq B \wedge \text{stab}(x) \in \Gamma \wedge \exists \xi < \alpha [\text{dom}(x) \subseteq V_\xi^{(\Gamma)}]\}.$$

We put

$$V^{(\Gamma)} = \{x : \exists \alpha (x \in V_\alpha^{(\Gamma)})\}.$$

It is now easy to verify that (*cf.* (1.6)):

$$x \in V^{(\Gamma)} \leftrightarrow \text{Fun}(x) \wedge \text{ran}(x) \subseteq B \wedge \text{dom}(x) \subseteq V^{(\Gamma)} \wedge \text{stab}(x) \in \Gamma,$$

and that

$$V^{(\Gamma)} \subseteq V^{(B)}.$$

For $u, v \in V^{(\Gamma)}$ we define $[\![u \in v]\!]^\Gamma$ and $[\![u = v]\!]^\Gamma$ recursively as we did $[\![u \in v]\!]^B$ and $[\![u = v]\!]^B$, that is,

$$[\![u \in v]\!]^\Gamma = \bigvee_{x \in \mathrm{dom}(v)} [v(x) \wedge [\![x = u]\!]^\Gamma],$$

$$[\![u = v]\!]^\Gamma = \bigwedge_{x \in \mathrm{dom}(u)} [u(x) \Rightarrow [\![x \in v]\!]^\Gamma] \wedge \bigwedge_{y \in \mathrm{dom}(v)} [v(y) \Rightarrow [\![x \in u]\!]^\Gamma].$$

It is then easily proved by induction that $[\![u \in v]\!]^\Gamma = [\![u \in v]\!]^B$, $[\![u = v]\!]^\Gamma = [\![u = v]\!]^B$, and so $[\![\cdot \in \cdot]\!]^\Gamma$, $[\![\cdot = \cdot]\!]^\Gamma$ turn $V^{(\Gamma)}$ into a B-valued structure. So if $\mathcal{L}^{(\Gamma)}$ is the language for $V^{(\Gamma)}$, that is, the result of expunging from $\mathcal{L}^{(B)}$ all constant symbols not denoting elements of $V^{(\Gamma)}$, the Boolean-value $[\![\sigma]\!]^\Gamma$ in $V^{(\Gamma)}$ of any $\mathcal{L}^{(\Gamma)}$-sentence σ is defined recursively by

$$[\![\sigma \wedge \tau]\!]^\Gamma = [\![\sigma]\!]^\Gamma \wedge [\![\tau]\!]^\Gamma$$

$$[\![\neg\sigma]\!]^\Gamma = ([\![\sigma]\!]^\Gamma)^*$$

$$[\![\exists x \phi(x)]\!]^\Gamma = \bigvee_{u \in V^{(\Gamma)}} [\![\phi(u)]\!]^\Gamma.$$

We shall need some technical facts about $V^{(\Gamma)}$.

Lemma 3.14 *For every* $x \in V$,

$$\hat{x} \in V^{(\Gamma)}.$$

Proof By induction on $\in$. Suppose $\hat{y} \in V^{(\Gamma)}$ for every $y \in x$. Then $\mathrm{dom}(\hat{x}) = \{\hat{y} : y \in x\} \subseteq V^{(\Gamma)}$ and by Theorem 3.3 we have $g\hat{x} = \hat{x}$ for very $g \in G$, whence $\mathrm{stab}(x) = G \in \Gamma$. Hence $\hat{x} \in V^{(\Gamma)}$. $\qquad\square$

From now on we assume that Γ is a normal filter of subgroups of G.

Lemma 3.15 G *acts on* $V^{(\Gamma)}$.

Proof One first shows by induction on $y \in \mathrm{dom}(x)$ that for any $g \in G$, the map $x \mapsto gx$ carries $V^{(\Gamma)}$ into $V^{(\Gamma)}$. Suppose then that $x \in V^{(\Gamma)}$ and $gy \in V^{(\Gamma)}$ for all $y \in \mathrm{dom}(x)$. Then $\mathrm{dom}(gx) = \{gy : y \in \mathrm{dom}(x)\} \subseteq V^{(\Gamma)}$. Also, it is readily verified that $\mathrm{stab}(gx) = g\,\mathrm{stab}(x)g^{-1}$, and therefore $\mathrm{stab}(gx) \in \Gamma$ by the normality of Γ. Hence $gx \in V^{(\Gamma)}$, completing the induction step.

Finally, since G acts on $V^{(B)}$, we have for any $g \in G, u, v \in V^{(\Gamma)}$,

$$g \cdot [\![u \in v]\!]^\Gamma = g \cdot [\![u \in v]\!]^B = [\![gu \in gv]\!]^B = [\![gu \in gv]\!]^\Gamma$$

and similarly for $u = v$. Therefore G acts on $V^{(\Gamma)}$ as claimed. $\qquad\square$

From Lemma 3.15 and (3.2) it follows that, for every formula $\phi(v_1, \ldots, v_n)$, all $x_1, \ldots, x_n \in V^{(\Gamma)}$ and all $g \in G$,

$$g \cdot [\![\phi(x_1, \ldots, x_n)]\!]^\Gamma = [\![\phi(gx_1, \ldots, gx_n)]\!]^\Gamma. \tag{3.16}$$

If P is a basis for B, we define, for any $p \in P$ and any $\mathcal{L}^{(\Gamma)}$-sentence σ, $p\,\Gamma$-*forces* σ by

$$p \Vdash_\Gamma \sigma \leftrightarrow p \leq [\![\sigma]\!]^\Gamma.$$

It follows immediately from (3.16) that for any formula $\phi(v_1, \ldots, v_n)$, any $x_1, \ldots, x_n \in V^{(\Gamma)}$ and any $p \in P, g \in G$ for which $gp \in P$,

$$p \Vdash_\Gamma \phi(x_1, \ldots, x_n) \to gp \Vdash_\Gamma \phi(gx_1, \ldots, gx_n). \tag{3.17}$$

We define the notions of *truth* and *validity* in $V^{(\Gamma)}$ in the same way as we defined those notions in $V^{(B)}$, for example, an $\mathcal{L}^{(\Gamma)}$-sentence σ is *true* in $V^{(\Gamma)}$ (and we write $V^{(\Gamma)} \models \sigma$) if $[\![\sigma]\!]^\Gamma = 1$.

The same argument as that used in the proof of Theorem 1.17 establishes the following.

Theorem 3.18 *Theorem 1.17 continues to hold when 'B' is replaced by 'Γ '.*

We can now show that $V^{(\Gamma)}$ is a Boolean-valued model of ZF.

Theorem 3.19 *All the axioms—and hence all the theorems—of ZF are true in* $V^{(\Gamma)}$.

Proof The axioms of extensionality and regularity go through as in Lemmas 1.34 and 1.42. As for the remaining axioms (with the exception of choice), the same proofs as those given in Lemmas 1.35–1.38, Definition 1.39, Problem 1.40, and Lemma 1.41 work, except that now one must verify that the object v required to exist by the axiom in question has its stabilizer $\mathrm{stab}(v)$ in Γ. We do this in detail for the axiom scheme of separation, confining ourselves to brief hints in the case of other axioms.

Separation Let $\psi(x, v_1, \ldots, v_n)$ be a formula, and let $u, a_1, \ldots, a_n \in V^{(\Gamma)}$ (thus the $a_1, \ldots, a_n$ are regarded as *parameters*). Define $v \in V^{(B)}$ by $\mathrm{dom}(v) = \mathrm{dom}(u)$ and

$$v(x) = u(x) \wedge [\![\psi(x, a_1, \ldots, a_n)]\!]^\Gamma$$

for $x \in \mathrm{dom}(v)$. It now suffices to show that $v \in V^{(\Gamma)}$, for then one readily verifies as in Lemma 1.35 that

$$V^{(\Gamma)} \models \forall x[x \in v \leftrightarrow x \in u \wedge \psi(x, a_1, \ldots, a_n)].$$

Since $\mathrm{dom}(v) = \mathrm{dom}(u) \subseteq V^{(\Gamma)}$, to show that $v \in V^{(\Gamma)}$ it is enough to prove that $\mathrm{stab}(v) \in \Gamma$. And since $\mathrm{stab}(u), \mathrm{stab}(a_1), \ldots, \mathrm{stab}(a_n)$ are all in Γ and Γ is a filter, it will be enough to show that

$$A = \mathrm{stab}(u) \cap \mathrm{stab}(a_1) \cap \cdots \cap \mathrm{stab}(a_n) \subseteq \mathrm{stab}(v). \qquad (*)$$

If $g \in A$, then $\mathrm{dom}(gv) = \{gx : x \in \mathrm{dom}(v)\} = \{gx : x \in \mathrm{dom}(u)\} = \mathrm{dom}(gu) = \mathrm{dom}(u) = \mathrm{dom}(v)$. Also, if $x \in \mathrm{dom}(v)$, then $x = gy$ with $y \in \mathrm{dom}(u)$ so that

$$
\begin{aligned}
(gv)(x) &= (gv)(gy) \\
&= g \cdot v(y) \\
&= g \cdot u(y) \wedge [\![\psi(gy, ga_1, \ldots, ga_n)]\!]^{\Gamma} \quad \text{(by (3.16))} \\
&= (gu)(gy) \wedge [\![\psi(x, a_1, \ldots, a_n)]\!]^{\Gamma} \\
&= u(x) \wedge [\![\psi(x, a_1, \ldots, a_n)]\!]^{\Gamma} \\
&= v(x).
\end{aligned}
$$

Hence $gv = v$ and $g \in \mathrm{stab}(v)$. This proves $(*)$ and the result in question.

Replacement In our original proof of the truth of this axiom in $V^{(B)}$ (Lemma 1.36) we used a set of the form $V_\alpha^{(B)} \times \{1\}$ to include the range of the 'function' defined by $\phi(x, y)$ on u. The same proof works here with $V_\alpha^{(B)} \times \{1\}$ replaced by $V_\alpha^{(\Gamma)} \times \{1\}$.

Union Given $u \in V^{(\Gamma)}$, define v by $\mathrm{dom}(v) = \bigcup\{\mathrm{dom}(y) : y \in \mathrm{dom}(u)\}$ and $v(x) = [\![\exists y \in u[x \in y]]\!]^{\Gamma}$. One then verifies that $\mathrm{stab}(u) \subseteq \mathrm{stab}(v)$, so that $v \in V^{(\Gamma)}$. As in the original verification of the truth of the union axiom in $V^{(B)}$ (Lemma 1.37) one shows that

$$V^{(\Gamma)} \models \forall x[x \in v \leftrightarrow \exists y \in u[x \in y]],$$

so that the axiom is true in $V^{(\Gamma)}$ as well.

Power Set. For $u \in V^{(\Gamma)}$ define V by $\mathrm{dom}(v) = B^{\mathrm{dom}(u)} \cap V^{(\Gamma)}$ and $v(x) = [\![x \subseteq u]\!]^{\Gamma}$ for $x \in \mathrm{dom}(v)$. One can now show that $\mathrm{stab}(u) \subseteq \mathrm{stab}(v)$, so that $v \in V^{(\Gamma)}$. As in the original proof of the truth of the power set axiom in $V^{(B)}$ (Lemma 1.38), one verifies that

$$V^{(\Gamma)} \models \forall x[x \in v \leftrightarrow x \subseteq u],$$

showing that the axiom is true in $V^{(\Gamma)}$ as well.

Infinity. We know that $\hat{\omega} \in V^{(\Gamma)}$ (Lemma 3.14), so the argument here is the same as in Lemma 1.41. $\qquad \square$

We now select specific B, G, and Γ in such a way that we have $V^{(\Gamma)} \models \neg AC$. It is clear that, if we can prove the existence of such B, G, and Γ in ZFC, then in view of Theorems 3.18 and 3.19, the same argument as we used to prove Theorem 1.19 implies that, if ZF is consistent, so is ZF $+ \neg$AC.

Let X be the product space $2^{\omega \times \omega}$ and let $B = \mathrm{RO}(2^{\omega \times \omega})$. Let G be the group of all permutations of ω. G can be made to act on B in the following way. Each $g \in G$ induces a homeomorphism g^* of X onto itself via

$$(g^* f)\langle m, n \rangle = f\langle m, gn \rangle$$

for $f \in X$ and $m, n \in \omega$. We define the action $\langle g, b \rangle \mapsto gb$ of G on B by

$$gb = g^{*-1}[b]$$
$$= \{ f \in X : g^* f \in b \}.$$

(It is readily checked that this does indeed define an action.)

For each $n \in \omega$ let $G_n = \{ g \in G : gn = n \}$; clearly this is a subgroup of G. Let Γ be the filter of subgroups generated by the G_n. That is, if for each finite subset $J \subseteq \omega$ we write

$$G_J = \bigcap_{n \in J} G_n$$

then Γ is the set of all subgroups H of G such that $G_J \subseteq H$ for some finite $J \subseteq \omega$. It is readily verified that Γ is normal.

Recalling that $P = C(\omega \times \omega, 2)$ is a basis for B, we now prove

Lemma 3.20 *If $p \in P$, J is a finite subset of ω and $n \notin J$, then there is $g \in G_J$ such that $p \wedge gp \neq 0$ and $gn \neq n$.*

Proof Take $n' \notin J \cup \{n\}$ so that $\langle m, n' \rangle \notin \mathrm{dom}(p)$ for any m (possible, since J and $\mathrm{dom}(p)$ are finite) and let $g \in G$ be the permutation of ω, which interchanges n and n' but leaves everything else undisturbed. Then certainly $g \in G_J$ and $gn \neq n$. To see that $p \wedge gp \neq 0$, recall that p is identified with the element

$$N(p) = \{ f \in 2^{\omega \times \omega} : p \subseteq f \}$$

of B and observe that

$$g \cdot N(p) = \{ f \in 2^{\omega \times \omega} : p \subseteq g^* f \}$$
$$= \{ f \in 2^{\omega \times \omega} : \langle i, j \rangle \in \mathrm{dom}(p) \to f\langle i, gj \rangle = p\langle i, j \rangle \}.$$

Let $i_1, \ldots, i_k$ be a list of the i such that $\langle i, n \rangle \in \mathrm{dom}(p)$. Then

$$p \wedge gp = N(p) \cap g \cdot N(p)$$
$$= \{ f \in 2^{\omega \times \omega} : p \subseteq f \text{ and } f\langle i_j, n' \rangle = p\langle i_j, n \rangle \text{ for } j = 1, \ldots, k \}$$
$$\neq \emptyset,$$

since $\langle i_j, n' \rangle \notin \mathrm{dom}(p)$ for $j = 1, \ldots, k$. $\qquad\square$

Now we can prove the following.

Theorem 3.21 *With B, G, and Γ as above, we have $V^{(\Gamma)} \models$ 'there is an infinite Dedekind finite subset of $P\hat{\omega}$' and so, a fortiori,*

$$V^{(\Gamma)} \models \neg \mathrm{AC}.$$

Proof We write $[\![\sigma]\!]$ for $[\![\sigma]\!]^{\Gamma}$ and $\Vdash$ for $\Vdash_{\Gamma}$ throughout. For each $n \in \omega$ define $u_n \in B^{\mathrm{dom}(\hat{\omega})}$ by

$$u_n(\hat{m}) = \{ h \in 2^{\omega \times \omega} : h\langle m, n \rangle = 1 \}.$$

The usual calculation shows that

$$V^{(B)} \models u_n \subseteq \hat{\omega}$$

for all $n \in \omega$. Moreover for all $g \in G$ and for all $n \in \omega$,

(1) $gu_n = u_{gn}$.

For clearly we have $\mathrm{dom}(gu_n) = \mathrm{dom}(u_{gn})$. Also, for $m \in \omega$,

$$(gu_n)\hat{m} = (gu_n)(g\hat{m})$$
$$= g \cdot u_n(\hat{m})$$
$$= g^{*-1}[\{ h \in 2^{\omega \times \omega} : h\langle m, n \rangle = 1 \}]$$
$$= \{ h \in 2^{\omega \times \omega} : g^* h\langle m, n \rangle = 1 \}$$
$$= \{ h \in 2^{\omega \times \omega} : h\langle m, gn \rangle = 1 \}$$
$$= u_{gn}(\hat{m}),$$

and (1) follows.

(1) immediately gives $G_n \subseteq \mathrm{stab}(u_n)$, so $\mathrm{stab}(u_n) \in \Gamma$ and therefore $u_n \in V^{(\Gamma)}$. The argument in the proof of Theorem 2.12 gives:

(2) $V^{(\Gamma)} \models u_n \neq u_{n'}$,   for $n \neq n'$.

Now put

$$s = \{ u_n : n \in w \} \times \{1\};$$

it is then easy to see that $gs = s$ for any $g \in G$, so $s \in V^{(\Gamma)}$. Moreover, it is not hard to verify that $V^{(\Gamma)} \models s \subseteq P\hat{\omega}$. It now follows from (2) that

$$V^{(\Gamma)} \models s \text{ is infinite.}$$

We claim that

$$V^{(\Gamma)} \models s \text{ is not Dedekind infinite,}$$

which will prove the theorem. To establish the claim, it suffices to show that, for each $f \in V^{(\Gamma)}$,

$$[\![\mathrm{Fun}(f) \wedge f \text{ is one--one} \wedge \mathrm{dom}(f) = \hat{\omega} \wedge \mathrm{ran}(f) \subseteq s]\!] = 0.$$

Suppose not; then there is $p_0 \in P = C(\omega \times \omega, 2)$ (the basis for B) such that

$$p_0 \Vdash \mathrm{Fun}(f) \wedge f \text{ is one--one} \wedge \mathrm{dom}(f) = \hat{\omega} \wedge \mathrm{ran}(f) \subseteq s.$$

We shall find $q \leq p_0$ such that

$$q \Vdash \neg\mathrm{Fun}(f),$$

which will yield the desired contradiction.

We first observe that,

(3) $p \Vdash x \in s \leftrightarrow \forall q \leq p \exists r \leq q \exists n \in \omega[r \Vdash x = u_n]$.

For we have

$$p \Vdash x \in s \leftrightarrow p \leq \bigvee_{n \in \omega} [\![x = u_n]\!]$$

$$\leftrightarrow p \wedge \bigwedge_{n \in \omega} [\![x \neq u_n]\!] = 0$$

$$\leftrightarrow \forall q \leq p \left[q \not\leq \bigwedge_{n \in \omega} [\![x \neq u_n]\!] \right]$$

$$\leftrightarrow \forall q \leq p \exists n \in \omega[q \not\leq [\![x \neq u_n]\!]]$$

$$\leftrightarrow \forall q \leq p \exists n \in \omega \neg[q \Vdash x \neq u_n]$$

$$\leftrightarrow \forall q \leq p \exists n \in \omega \exists r \leq q[r \Vdash x = u_n].$$

Now since $f \in V^{(\Gamma)}$ it has a finite support J, that is, there is a finite subset $J \subseteq \omega$ such that $G_J \subseteq \mathrm{stab}(f)$. Let

$$J = \{n_1, \ldots, n_j\}.$$

Since $p_0 \Vdash f$ *is one–one* $\wedge\mathrm{Fun}(f)$, it follows that

$$p_0 \Vdash \exists x \in \hat{\omega}[f(x) \neq u_{n_1} \wedge \cdots \wedge f(x) \neq u_{n_j}]$$

so that there is $p \leq p_0$ and $m \in w$ such that,

(4) $p \Vdash f(\hat{m}) \neq u_{n_1} \wedge \cdots \wedge f(\hat{m}) \neq u_{n_j}$.

Since $p_0 \Vdash f(\hat{m}) \in s$, so that $p \Vdash f(\hat{m}) \in s$, by (3) there is $r \leq p$ and $n \in \omega$ such that,

(5) $r \Vdash f(\hat{m}) = u_n$.

But (4) implies

$$r \Vdash f(\hat{m}) \neq u_{n_1} \wedge \cdots \wedge f(\hat{m}) \neq u_{n_j}$$

and this, together with (5), implies $n \neq n_1 \wedge \cdots \wedge n \neq n_j$, that is, $n \notin J$. By Lemma 3.20 there is $g \in G_J$ such that $p \wedge gp = 0$ and $gn \neq n$. It follows from (5) and (3.17) that

$$gr \Vdash (gf)g\hat{m} = gu_n.$$

But this, together with (1) and the fact that $g \in G_J \subseteq \mathrm{stab}(f)$ gives

$$gr \Vdash f(\hat{m}) = u_{gn}.$$

Since $r \wedge gr \neq 0$, there is $q \in P$ such that $q \leq r$ and $q \leq gr$. Then $q \leq p_0$ and

$$q \Vdash f(\hat{m}) = u_n \wedge f(\hat{m}) = u_{gn}.$$

But since $gn \neq n$, we have, using (2), $[\![u_{gn} \neq u_n]\!] = 1$, so that $q \Vdash u_{gn} \neq u_n$. Therefore

$$q \Vdash \neg\mathrm{Fun}(f),$$

and the proof is complete. $\qquad\square$

Corollary 3.22 *If* ZF *is consistent, so is* ZF $+ \neg$AC.

We concluded this chapter with some remarks on the origins of the proof of Theorem 3.21. The construction of $V^{(\Gamma)}$ is in fact derived from an earlier construction, due to Fraenkel and Mostowski, which was used to show that AC is independent of a certain modified form of ZF, namely, the theory ZFA of set theory with atoms. (This is actually a weaker result because the axiom of foundation does not hold in ZFA.) To obtain ZFA from ZF, one drops the axiom of foundation, and adds an axiom asserting the existence of a nonempty set A

of *atoms*, that is, objects lacking members, yet not identical with the empty set. The axiom of extensionality must also be suitably modified. The method of Fraenkel–Mostowski now runs roughly as follows. One first shows that the permutation group G of A acts on the universe V by $\in$-automorphisms. Then, letting Γ be a normal filter in G, one constructs $V^{(\Gamma)}$ essentially as before and shows that it is a model of ZFA. By choosing A and Γ properly, one can arrange things so that in $V^{(\Gamma)}$ the set A is not well-orderable. This will be so essentially because the presence in $V^{(\Gamma)}$ of so many automorphisms permuting the elements of A will make these elements effectively indiscernible in $V^{(\Gamma)}$ and thereby render it impossible to chooe a 'first element' of A. (For more on Fraenkel–Mostowski models, see Felgner (1971) or Jech (1973).)

The method we have presented for proving the independence of AC from ZF combines the Fraenkel–Mostowski technique with that of Boolean-valued models. In a nutshell, a set of 'reals' in $V^{(\Gamma)}$ (the set $\{u_n : n \in \omega\}$) is found, which behaves very much like a set of atoms; then one argues à la Fraenkel–Mostowski. It turns out that many Fraenkel–Mostowski proofs of independence from ZFA can be converted in this way into proofs of independence from ZF, for example, those of 'every set can be linearly ordered', 'every countable set of pairs has a choice function', 'every vector space has a basis'. See Jech (1973).

4

GENERIC ULTRAFILTERS AND TRANSITIVE MODELS OF ZFC

In this chapter we replace V by a (transitive) model M of ZFC such that $M \in V$, and perform the construction of $V^{(B)}$ and $[\![\cdot]\!]^B$ *inside M*. We shall see that this construction gives rise to models of ZFC in which one can falsify the various set-theoretic assertions whose formal independence of ZFC was established in earlier chapters. In this way Boolean-valued set theory can be transformed into a valuable model-theoretic tool.

Now let M be a transitive $\in$-model of ZFC and let $B \in M$ be a complete Boolean algebra in the sense of M. That is,

$$M \models B \text{ is a complete Boolean algebra.}$$

In particular, if $X \in PB \cap M$, then $\bigvee X$ and $\bigwedge X$ exist and are in M. (Notice that $PB \cap M$ is the power set of B formed in M, $P^{(M)}(B)$.) Moreover, since the predicate 'B is a Boolean algebra' is a restricted formula, it follows that B is a Boolean algebra (but not necessarily complete) 'from the outside' as well.

Under these conditions we can relativize all the notions and constructions of Chapter 1 to M. We write $M^{(B)}$ for $(V^{(B)})^{(M)}$ and $\mathcal{L}_M^{(B)}$ for $(\mathcal{L}^{(B)})^{(M)}$. $M^{(B)}$ is called the *B-extension* of M: all the results proved in Chapter 1 for $V^{(B)}$ hold, *mutatis mutandis*, for $M^{(B)}$. We also obtain a Boolean truth value $([\![\sigma]\!]^B)^{(M)} \in B$ for each $\mathcal{L}_M^{(B)}$-sentence σ: to simplify the notation we agree to write $[\![\sigma]\!]$ for $([\![\sigma]\!]^B)^{(M)}$. Writing $M^{(B)} \models \sigma$ for $[\![\sigma]\!] = 1$, we see from Theorem 1.33 that $M^{(B)} \models \sigma$ whenever σ is an axiom of ZFC, so that $M^{(B)}$ may be thought of as a Boolean-valued *model* of ZFC. Finally, by analogy with Definition 1.22, we obtain a map ˆ: $M \to M^B$ such that, for each $x \in M$,

$$\hat{x} = \{\langle \hat{y}, 1 \rangle : y \in x\}.$$

Now suppose that U is an arbitrary but fixed *ultrafilter* in B. In general, U is *not* a member of M. Define the relation $\sim_U$ on $M^{(B)}$ by putting, for $x, y \in M^{(B)}$,

$$x \sim_U y \leftrightarrow [\![x = y]\!] \in U.$$

It is easy to verify that $\sim_U$ is an *equivalence relation* on $M^{(B)}$. For $x \in M^{(B)}$, write x^U for the $\sim_U$-class of x, that is,

$$x^U = \{y \in M^{(B)} : x \sim_U y\},$$

and define the relation $\in_U$ on the set of all $\sim_U$-classes by

$$x^U \in_U y^U \leftrightarrow [\![x \in y]\!] \in U.$$

Now define the *quotient* of $M^{(B)}$ by U to be the structure

$$M^{(B)}/U = \langle \{x^U : x \in M^{(B)}\}, \in_U \rangle.$$

Theorem 4.1 *For any formula* $\phi(v_1, \ldots, v_n)$ *and any* $x_1, \ldots, x_n \in M^{(B)}$,

$$M^{(B)}/U \models \phi[x_1^U, \ldots, x_n^U] \leftrightarrow [\![\phi(x_1, \ldots, x_n)]\!] \in U.$$

Proof Induction on the complexity of ϕ, using the Maximum Principle to handle the existential case. We omit the straightforward details. $\qquad\square$

Corollary 4.2 $M^{(B)}/U$ *is a model of* ZFC. *More generally, for any sentence* σ, *if* $M^{(B)} \models \sigma$, *then* $M^{(B)}/U \models \sigma$.

Let $S \subseteq P^{(M)}(B) = PB \cap M$. Recall (Chapter 0) that U is said to be S-complete if, for all $X \in S$,

$$\bigvee X \in U \to X \cap U \neq \emptyset$$

(the reverse implication holding trivially). A $P^{(M)}(B)$-complete ultrafilter is called *M-generic*.

A partition of unity $\{a_i : i \in I\}$ in B is called an *M-partition of unity in B* if $\langle a_i : i \in I \rangle \in M$. We have the following simple characterization of M-genericity in terms of this notion:

Lemma 4.3 *The following conditions are equivalent:*

(i) U *is M-generic;*
(ii) *for any M-partition of unity* $\{a_i : i \in I\}$ *in B, there is* $i \in I$ *such that* $a_i \in U$.

Proof (i) $\to$ (ii) is clear. Conversely, assume (ii) and let $A \in P^{(M)}(B)$. Since the axiom of choice holds in M, there is an ordinal $\alpha \in M$ such that $A \cup \{(\bigvee A)^*\} = \{a_\xi : \xi < \alpha\}$. Now put $b_\xi = a_\xi - \bigvee_{\eta < \xi} a_\eta$ for $\xi < \alpha$. Then $\{b_\xi : \xi < \alpha\}$ is an M-partition of unity in B and so by (ii) there is $\xi < \alpha$ such that $b_\xi \in U$. It is now easy to see that $\bigvee A \in U \to U \cap A \neq \emptyset$. Hence U is a M-generic. $\qquad\square$

An element $\alpha \in M^{(B)}/U$ is called an *ordinal in* $M^{(B)}/U$ if $M^{(B)}/U \models \mathrm{Ord}[\alpha]$. It follows immediately from Theorem 4.1 that if α is an ordinal in M then $\hat{\alpha}^U$ is an ordinal in $M^{(B)}/U$; elements of the latter form are called *standard* ordinals in $M^{(B)}/U$.

Let $\mathrm{ORD}^{(M)}$ be the set of all ordinals in M (thus $\mathrm{ORD}^{(M)} = \mathrm{ORD} \cap M$). Then for $x \in M^{(B)}$ the set

$$\{ [\![x = \hat{\alpha}]\!] : \alpha \in \mathrm{ORD}^{(M)} \} \tag{4.4}$$

is a subset of B which is definable in M (from the parameter x) and is therefore a member of M. Let S_1 be the subfamily of $P^{(M)}(B)$ consisting of all sets of the form (4.4) for $x \in M^{(B)}$. By Theorem 1.44 we have, for $x \in M^{(B)}$,

$$[\![\mathrm{Ord}(x)]\!] = \bigvee_{\alpha \in \mathrm{ORD}^{(M)}} [\![x = \hat{\alpha}]\!]. \tag{4.5}$$

We use this in the proof of

Theorem 4.6 *The following conditions are equivalent:*

(i) *U is S_1-complete;*
(ii) *all ordinals in $M^{(B)}/U$ are standard;*
(iii) *U is M-generic.*

Proof (i) $\rightarrow$ (iii) Assume (i) and let $\{ a_\xi : \xi < \alpha \}$ be an M-partition of unity in B, where $\alpha \in \mathrm{ORD}^{(M)}$. (Since the axiom of choice holds in M there is no loss of generality in assuming the partition of unity to be indexed by an ordinal α.) By Problem 1.26(i), there is $x \in M^{(B)}$ such that $a_\xi = [\![x = \hat{\xi}]\!]$ for $\xi < \alpha$. We have

$$\bigvee_{\xi \in \mathrm{ORD}^{(M)}} [\![x = \hat{\xi}]\!] \geq \bigvee_{\xi < \alpha} [\![x = \hat{\xi}]\!] = \bigvee_{\xi < \alpha} a_\xi = 1 \in U,$$

so that, by (i), there is $\eta \in \mathrm{ORD}^{(M)}$ such that $[\![x = \hat{\eta}]\!] \in U$. If $\eta \geq \alpha$, then

$$[\![x = \hat{\eta}]\!] = [\![x = \hat{\eta}]\!] \wedge 1 = [\![x = \hat{\eta}]\!] \wedge \bigvee_{\xi < \alpha} [\![x = \hat{\xi}]\!] = 0 \notin U,$$

so that we must have $\eta < \alpha$. Hence $a_\eta = [\![x = \hat{\eta}]\!] \in U$ and (iii) follows by Lemma 4.3.

(iii) $\to$ (ii) Assume (iii); then, using (4.5) and Theorem 4.1,

$$M^{(B)}/U \models \mathrm{Ord}[x^U] \leftrightarrow [\![\mathrm{Ord}(x)]\!] \in U$$

$$\leftrightarrow \bigvee_{\alpha \in \mathrm{ORD}^{(M)}} [\![x = \hat{\alpha}]\!] \in U$$

$$\leftrightarrow [\![x = \hat{\alpha}]\!] \in U \text{ for some } \alpha \in \mathrm{ORD}^{(M)}$$

$$\leftrightarrow x^U = \hat{\alpha}^U \text{ for some } \alpha \in \mathrm{ORD}^{(M)}.$$

(ii) $\to$ (i) Assume (ii); then if $\bigvee_{\alpha \in \mathrm{ORD}^{(M)}} [\![x = \hat{\alpha}]\!] \in U$, we have $[\![\mathrm{Ord}(x)]\!] \in U$ by (4.5), so $M^{(B)}/U \models \mathrm{Ord}[x^U]$ by Theorem 4.1. Hence (ii) gives $x^U = \hat{\alpha}^U$ for some $\alpha \in \mathrm{ORD}^{(M)}$, whence $[\![x = \hat{\alpha}]\!] \in U$. $\qquad\square$

Corollary 4.7[1] *If U is M-generic, then $\in_U$ is a well-founded relation.*

Proof If U is M-generic, then Theorem 4.6 implies that the map $\alpha \mapsto \hat{\alpha}^U$ sends the well-ordered set $\mathrm{ORD}^{(M)}$ onto the set of ordinals in $M^{(B)}/U$. This map is easily seen to be order-preserving (with respect to $\in_U$) and it follows that the ordinals of $M^{(B)}/U$ are well-ordered by $\in_U$. The usual rank argument now shows that $\in_U$ is well-founded on $M^{(B)}/U$: if not, then there would be an infinite descending $\in_U$-sequence $\ldots x_2 \in_U x_1 \in_U x_0$; if ρ is the rank function in $M^{(B)}/U$, then $\ldots, \rho(x_2), \rho(x_1), \rho(x_0)$ would be an infinite descending sequence of ordinals in $M^{(B)}/U$, contradicting the fact that these are well-ordered. $\qquad\square$

Suppose now that $\in_U$ is a well-founded relation. Then, by Mostowski's collapsing lemma, $M^{(B)}/U$ can be collapsed to a unique transitive $\in$-structure $M[U]$ via the map h defined recursively on $\in_U$ by

$$h(x^U) = \{h(y^U) : y^U \in_U x^U\} = \{h(y^U) : [\![y \in x]\!] \in U\}. \qquad (4.8)$$

Thus $h : M^{(B)}/U \to M[U]$ is a bijection satisfying

$$x^U \in_U y^U \leftrightarrow h(x^U) \in h(y^U).$$

We can now define a map i of $M^{(B)}$ onto $M[U]$ by putting

$$i(x) = h(x^U) \qquad (4.9)$$

for $x \in M^{(B)}$. By (4.8) we have, for $x \in M^{(B)}$,

$$i(x) = \{i(y) : [\![y \in x]\!] \in U\}. \qquad (4.10)$$

---

[1]The converse of Corollary 4.7 fails: see Problem 4.32.

The map i—which we sometimes write as i_U—is called the *canonical map of* $M^{(B)}$ *onto* $M[U]$.

Lemma 4.11 *For any formula* $\phi(v_1, \ldots, v_n)$ *and any* $x_1, \ldots, x_n \in M^{(B)}$,

$$M[U] \models \phi[i(x_1), \ldots, i(x_n)] \leftrightarrow [\![\phi(x_1, \ldots, x_n)]\!] \in U.$$

Proof Immediate from Theorem 4.1 and the fact that h is an isomorphism of $M^{(B)}/U$ onto $M[U]$. $\qquad\qquad\square$

Now define $j : M \to M[U]$ by

$$j(x) = i(\hat{x}) \tag{4.12}$$

for $x \in M$. We see immediately from Theorem 1.23 that j is a one–one map satisfying $x \in y \leftrightarrow j(x) \in j(y)$; that is, j is an $\in$-monomorphism of M into $M[U]$. The situation can be depicted by the commutative diagram:

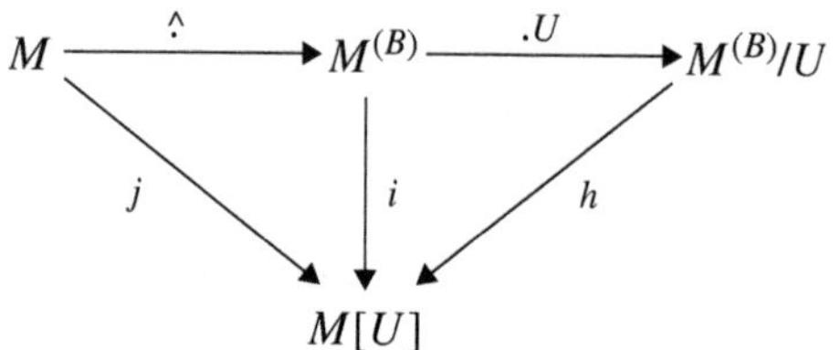

If $x \in M^{(B)}, y \in M$, the set

$$\{[\![x = \hat{z}]\!] : z \in y\} \tag{4.13}$$

is a subset of B which is definable in M (from the parameters x, y) and is accordingly a member of M. Let S_2 be the subfamily of $P^{(M)}(B)$ consisting of all sets of the form (4.13) for $x \in M^{(B)}, y \in M$. We recall from Theorem 1.23(i) that, for $x \in M^{(B)}, y \in M$,

$$[\![x \in \hat{y}]\!] = \bigvee_{z \in y} [\![x = \hat{z}]\!]. \tag{4.14}$$

We use this in the proof of the following.

Theorem 4.15 *The following conditions are equivalent:*

   (i) U *is* S_2*-complete;*
   (ii) $\in_U$ *is well-founded and* j *is the identity on* M;
   (iii) $\in_U$ *is well-founded and* $j[M]$ *is transitive;*
   (iv) U *is* M*-generic.*

Proof The equivalence of (ii) and (iii) follows easily from the transitivity of M and the fact that, by construction, j is an $\in$-isomorphism of M onto $j[M]$.

(i) $\to$ (iv). Assume (i), and let $\{a_i : i \in I\}$ be an M-partition of unity in B. By Problem 1.26(i) there is $x \in M^{(B)}$ such that $a_i = [\![x = \hat{a}_i]\!]$ for $i \in I$. Putting $\{a_i : i \in I\} = y \in M$, we have

$$\bigvee_{z \in y} [\![x = \hat{z}]\!] = \bigvee_{i \in I} [\![x = \hat{a}_i]\!] = \bigvee_{i \in I} a_i = 1 \in U.$$

Therefore, by (i), there is $z \in y$ such that $[\![x = \hat{z}]\!] \in U$. But $z = a_i$ for some $i \in I$, whence $a_i = [\![x = \hat{a}_i]\!] \in U$. (iv) now follows from Lemma 4.3.

(iv) $\to$ (iii). Assume (iv). Then $\in_U$ is well-founded by Corollary 4.7. Also, if $y \in M$ and $x \in j(y)$, then, since $M[U]$ is transitive, there is $x' \in M^{(B)}$ such that $x = i(x')$. Thus $i(x') \in j(y) = i(\hat{y})$, so that, by (4.10), $[\![x' \in \hat{y}]\!] \in U$. It follows now from (iv) and (4.14) that there is $z \in y$, hence $z \in M$, such that $[\![x' = \hat{z}]\!] \in U$, whence $x = j(z) \in j[M]$, and (iii) follows.

(iii) $\to$ (i). Assume (iii). If $x \in M^{(B)}, y \in M$, then

$$\bigvee_{z \in y} [\![x = \hat{z}]\!] \in U \to [\![x \in \hat{y}]\!] \in U \qquad \text{(by (4.14))}$$

$$\to x^U \in_U \hat{y}^U$$
$$\to h(x^U) \in h(\hat{y}^U)$$
$$\to i(x) \in i(\hat{y}) = j(y)$$
$$\to i(x) = j(z) \text{ for some } z \in M \qquad \text{(by (iii))}$$
$$\to j(z) \in j(y)$$
$$\to z \in y.$$

Hence $i(x) = j(z) = i(\hat{z})$, so by definition of i (4.9), $[\![x = \hat{z}]\!] \in U$. (i) follows. $\qquad \square$

Corollary 4.16 *If U is M-generic, then $M \subseteq M[U]$.*

Next, we define $U_* \in M^{(B)}$ by $\text{dom}(U_*) = \{\hat{x} : x \in B\} = \text{dom}(\hat{B})$ and $U_*(\hat{x}) = x$ for $x \in B$. Notice that U_* does *not* depend on U. We have for $x \in M^{(B)}$,

$$[\![x \in U_*]\!] = \bigvee_{y \in B} [y \wedge [\![x = \hat{y}]\!]], \qquad (4.17)$$

and, for $x \in B$,

$$[\![\hat{x} \in U_*]\!] = x. \qquad (4.18)$$

In particular, if σ is any B-sentence, we have $[\![[\![\sigma]\!]\,\hat{}\; \in U_*]\!] = [\![\sigma]\!]$, that is,

$$M^{(B)} \models [[\![\sigma]\!]\,\hat{}\; \in U_* \leftrightarrow \sigma].$$

In other words, U_* represents the truth values of 'true' sentences in $M^{(B)}$.

Let S_3 be the subfamily of $P^{(M)}(B)$ consisting of all subsets of B of the form $\{y \wedge [\![x = \hat{y}]\!] : y \in B\}$ for $x \in M^{(B)}$. Then we have

Theorem 4.19 *The following conditions are equivalent:*

   (i) *U is S_3-complete;*

  (ii) *$\in_U$ is well-founded, j is the identity on U, and $i(U_*) = U$;*

 (iii) *U is M-generic.*

Proof (i) $\rightarrow$ (iii). Assume (i). Let $\{a_i : i \in I\}$ be an M-partition of unity in B. By Problem 1.26(i) there is $x \in M^{(B)}$ such that $a_i = [\![x = \hat{a}_i]\!]$ for $i \in I$. Hence

$$\bigvee_{y \in B} [y \wedge [\![x = \hat{y}]\!]] \geq \bigvee_{i \in I} [a_i \wedge [\![x = \hat{a}_i]\!]] = \bigvee_{i \in I} a_i = 1 \in U,$$

and so by (i) there is $y \in B$ such that $y \wedge [\![x = \hat{y}]\!] \in U$. If $y \notin \{a_i : i \in I\}$ then

$$[\![x = \hat{y}]\!] = [\![x = \hat{y}]\!] \wedge 1 = [\![x = \hat{y}]\!] \wedge \bigvee_{i \in I} [\![x = \hat{a}_i]\!] = 0 \notin U,$$

so that there must be $i \in I$ such that $y = a_i$. But then $a_i = a_i \wedge [\![x = \hat{a}_i]\!] \in U$ and so (iii) follows by Lemma 4.3.

   (iii) $\rightarrow$ (ii). Assume (iii). Then by Theorem 4.15 $\in_U$ is well-founded and j is the identity on M, hence on U. We claim that, for $x \in M^{(B)}$,

$$\{i(x) : [\![x \in U_*]\!] \in U\} = \{i(\hat{y}) : y \in U\}. \tag{1}$$

This follows from the chain of equivalences

$$[\![x \in U_*]\!] \in U \leftrightarrow \bigvee_{y \in B} [y \wedge [\![x = \hat{y}]\!]] \in U \qquad \text{(by (4.17))}$$

$$\leftrightarrow \exists y \in B[y \wedge [\![x = \hat{y}]\!] \in U] \qquad \text{(by (iii))}$$

$$\leftrightarrow \exists y \in U[[\![x = \hat{y}]\!] \in U]$$

$$\leftrightarrow \exists y \in U[i(x) = i(\hat{y})].$$

Now we have

$$
\begin{aligned}
i(U_*) &= \{i(x) : [\![x \in U_*]\!] \in U\} && \text{(by (4.10))} \\
&= \{i(\hat{y}) : y \in U\} && \text{(by (1))} \\
&= \{j(y) : y \in U\} \\
&= \{y : y \in U\} = U,
\end{aligned}
$$

and this gives (ii).

(ii) $\rightarrow$ (i). Assuming (ii) we have, using (4.10)

$$
U = i(U_*) = \{i(x) : [\![x \in U_*]\!] \in U\}. \tag{2}
$$

Hence

$$
\begin{aligned}
\bigvee_{y \in B} [y \wedge [\![x = \hat{y}]\!]] \in U &\rightarrow [\![x \in U_*]\!] \in U && \text{(by (4.17))} \\
&\rightarrow i(x) \in U && \text{(by (2))} \\
&\rightarrow i(x) = y = j(y) \text{ for some } y \in U && \text{(by (ii))} \\
&\rightarrow i(x) = i(\hat{y}) \\
&\rightarrow [\![x = \hat{y}]\!] \in U \\
&\rightarrow y \wedge [\![x = \hat{y}]\!] \in U
\end{aligned}
$$

and (i) follows. $\qquad\square$

Theorem 4.19 immediately gives the following.

Corollary 4.20 *If U is M-generic, then $U \in M[U]$.*

We now show that U_* is an 'M-generic ultrafilter' in the sense of $M^{(B)}$. Given a Boolean algebra A and a subset $F \subseteq PA$, we say that A is F-*complete* if for all $X \in F, \bigvee X$ and $\bigwedge X$ exist in A. Thus our given Boolean algebra B is $P^{(M)}(B)$-complete. Since the predicate 'A is an F-complete Boolean algebra' is clearly a restricted formula (with parameters A and F), it follows that, using Theorem 1.23,

$$
M^{(B)} \models \hat{B} \text{ is a } (P^{(M)}(B))\,\hat{}\text{-complete Boolean algebra,}
$$

and furthermore we have the following.

Theorem 4.21 $M^{(B)} \models U_*$ *is a* $(P^{(M)}(B))\hat{}$ *-complete ultrafilter in* $\hat{B}$.

Proof Let us put $C = \hat{B}$. Then we know that

$$M^{(B)} \models C \text{ is a } (P^{(M)}(B))\hat{} \text{-complete Boolean algebra.}$$

Let us denote the Boolean operations in C by $\wedge_C, {}^{*c}, \vee_C$, etc. and the natural partial ordering in C by $\leq_C$. It is then easy to see that, for any $a, b \in B, A \subseteq B$, we have

$$[\![\hat{a} \wedge_C \hat{b} = (a \wedge b)\hat{}\,]\!] = [\![\hat{a}^{*C} = (a^*)\hat{}\,]\!] = 1$$

$$\left[\!\!\left[\bigvee_C \hat{A} = \left(\bigvee A \right)\hat{}\, \right]\!\!\right] = 1.$$

We turn now to the properties of U_*. To prove the theorem it will be enough to verify the assertions (a)–(e) below.

(a) $[\![U_* \subseteq C \wedge \hat{0} \notin U_*]\!] = 1$. This follows in a straightforward way from (4.17) and (4.18).

(b) $[\![\forall xy \in C[x, y \in U_* \to x \wedge_C y \in U_*]]\!] = 1$. To verify this, notice that the l.h.s. is

$$\bigwedge_{a,b \in B} [[\![\hat{a} \in U_* \wedge \hat{b} \in U_*]\!] \Rightarrow [\![\hat{a} \wedge_C \hat{b} \in U_*]\!]]$$

$$= \bigwedge_{a,b \in B} [[\![\hat{a} \in U_* \wedge \hat{b} \in U_*]\!] \Rightarrow (a \wedge b)\hat{} \in U_*]\!]]$$

$$= \bigwedge_{a,b \in B} [a \wedge b \Rightarrow a \wedge b] = 1.$$

(c) $[\![\forall xy \in C[y \in U_* \wedge y \leq_C x \to x \in U_*]]\!] = 1$. The proof of this is similar to (b), and is left to the reader.

(d) $[\![\forall x \in C[x \in U_* \vee x^{*C} \in U_*]]\!]$. The proof of this is also similar to (b).

(e) $[\![U_* \text{ is } (P^{(M)}(B))\hat{}\text{-complete}]\!] = 1$. Now the l.h.s. here is

$$\left[\!\!\left[\forall X \in (P^{(M)}(B))\hat{} \left[\bigvee X \in U_* \to U_* \cap X \neq \emptyset \right] \right]\!\!\right]$$

$$= \bigwedge_{A \in P^{(M)}(B)} \left[\!\!\left[\bigvee_C \hat{A} \in U_* \to U_* \cap \hat{A} \neq \emptyset \right]\!\!\right]$$

and this last expression $= 1$, in view of the fact that, for any $A \in P^{(M)}(B)$,

$$\left[\!\left[\bigvee_C \hat{A} \in U_* \right]\!\right] = \left[\!\left[\left(\bigvee A \right)^{\widehat{}} \in U_* \right]\!\right]$$

$$= \bigvee A$$

$$= \bigvee_{a \in A} [\![\hat{a} \in U_*]\!]$$

$$= [\![\exists x \in \hat{A}(x \in U_*)]\!]$$

$$= [\![U_* \cap \hat{A} \neq \emptyset]\!]. \qquad \square$$

In view of Theorem 4.21, U_* is called the *canonical generic ultrafilter* in $M^{(B)}$ (or in $\hat{B}$): if U is a generic ultrafilter in B, we see from Theorem 4.19 that U_* is the natural preimage of U under the map $i_U : M^{(B)} \to M[U]$.

If U is an M-generic ultrafilter in B, $M[U]$ is called a *generic extension* of M. We can now give an *invariant* characterization of $M[U]$ for generic U.

Theorem 4.22 *Let U be an M-generic ultrafilter in B. Then:*

  (i) $M[U]$ *is a transitive* $\in$-*model of* ZFC;
  (ii) $M[U]$ *is the least transitive* $\in$-*model of* ZF *which includes* M *and contains* U;
  (iii) M *and* $M[U]$ *have the same ordinals and constructible sets.*

Proof (i) Since $M[U]$ is, by construction, isomorphic to $M^{(B)}/U$, (i) is an immediate consequence of Corollary 4.2.

(ii) Let N be a transitive $\in$-model of ZF such that $M \subseteq N$ and $U \in N$. For each $\alpha \in \mathrm{ORD}^{(M)}$, put $M_\alpha^{(B)} = (V_\alpha^{(B)})^{(M)}$; then clearly $M_\alpha^{(B)} \in M \subseteq N$ and $M^{(B)} = \bigcup \{ M_\alpha^{(B)} : \alpha \in \mathrm{ORD}^{(M)} \}$. Inspection of the definition of the collapsing map i reveals that $i | M_\alpha^{(B)}$, the restriction of i to $M_\alpha^{(B)}$, can be defined in N from U; it follows that $\mathrm{ran}(i | M_\alpha^{(B)}) \in N$, and, since N is transitive, that $\mathrm{ran}(i | M_\alpha^{(B)}) \subseteq N$. Hence

$$M[U] = \bigcup \{ \mathrm{ran}(i | M_\alpha^{(B)} : \alpha \in \mathrm{ORD}^{(M)} \} \subseteq N.$$

This proves (ii).

(iii) Let $y \in M[U]$; then $y = i(x)$ for some $x \in M^{(B)}$ and we have

$$M[U] \models \mathrm{Ord}[y]$$

$$\leftrightarrow M[U] \models \mathrm{Ord}[i(x)]$$

$$\leftrightarrow [\![\mathrm{Ord}(x)]\!] \in U \qquad\qquad \text{(by Lemma 4.11)}$$

$$\leftrightarrow \bigvee_{\alpha \in \mathrm{ORD}^{(M)}} [\![x = \hat{\alpha}]\!] \in U \qquad\qquad \text{(by (4.5))}$$

$$\leftrightarrow \exists \alpha \in \mathrm{ORD}^{(M)} [[\![x = \hat{\alpha}]\!] \in U] \qquad\qquad \text{(since U is generic)}$$

$$\leftrightarrow \exists \alpha \in \mathrm{ORD}^{(M)} [y = i(x) = i(\hat{\alpha}) = \alpha] \qquad\qquad \text{(by Theorem 4.15)}$$

$$\leftrightarrow y \in \mathrm{ORD}^{(M)}.$$

Thus M and $M[U]$ have the same ordinals. That M and $M[U]$ have the same constructible sets is proved similarly, now using Theorem 1.46 (in M). $\qquad\square$

In view of (ii) of this theorem, $M[U]$ may be termed the model of ZFC obtained by *adjoining* U to M, or *generated* by U and M.

Under what conditions do M-generic ultrafilters exist? The following simple example shows that if M is *uncountable* one cannot in general establish the existence of generic ultrafilters. Let M be an uncountable transitive $\in$-model of ZFC such that $\omega_1 \in M$, and put $B = \mathrm{RO}(\omega_1^\omega)^{(M)}$. By Corollary 5.2 to be proved in Chapter 5, $M^{(B)} \models \hat{\omega}_1$ *is countable*. So if U were an M-generic ultrafilter in B, we would have $M[U] \models \omega_1$ *is countable* and hence, since $\omega_1 \in M \subseteq M[U], \omega_1$ would also be countable in the real world. This contradiction shows that there are no M-generic ultrafilters in B.

The situation is quite different, however, when M is *countable*.

Theorem 4.23 *If M is countable, (or, more generally, if $P^{(M)}(B)$ is countable) then for each $b \neq 0$ in B there is an M-generic ultrafilter in B containing b.*

Proof If M is countable, then so is $P^{(M)}(B)$, and the Rasiowa–Sikorski theorem (Theorem 0.6) applies. $\qquad\square$

Corollary 4.24 *Let σ be any sentence and suppose that in $\mathcal{L}$ we can define a constant term t such that*

$$\mathrm{ZF} + V = L \vdash [t \text{ is a complete Boolean algebra and } V^{(t)} \models \sigma].$$

Then, given any countable transitive $\in$-model of ZF, we can construct a countable transitive $\in$-model of $\mathrm{ZFC} + \sigma$.

Proof Let N be a countable transitive model of ZF, and let M be the submodel of N consisting of all members of N which are constructible in N, that is,

$M = \{x \in N : N \models L[x]\}$. It is well-known that M is then a transitive model of $ZF + V = L$. Let $B = t^{(M)}$; then B is a complete Boolean algebra in the sense of M and $M^{(B)} \models \sigma$. Let U be an M-generic ultrafilter in B, which exists by Theorem 4.23. Then $[\![\sigma]\!]^B = 1 \in U$, so that $M[U] \models \sigma$, and $M[U]$ is the required model. $\qquad\square$

It follows immediately from this corollary and the results of Chapter 2 that, given a countable transitive $\in$-model M of ZF, we can construct countable transitive $\in$-models of, for example,

$$ZFC + GCH + P\omega \nsubseteq L \qquad \text{(Theorem 2.8, 2.6)},$$

$$ZFC + 2^{\aleph_0} = \aleph_2 + \forall \kappa \geq \aleph_1 [2^\kappa = \kappa^+] \qquad \text{(Problem 2.19)},$$

$$ZFC + 2^{\aleph_0} = \aleph_1 + 2^{\aleph_1} = \aleph_{\omega+1} \qquad \text{(Problem 2.20)},$$

$$ZFC + GCH + P\omega \subseteq L + P\omega_1 \nsubseteq L \qquad \text{(Problem 2.21)}.$$

We conclude this chapter by showing how results about $M^{(B)}$ can be 'transferred' to $V^{(B)}$. The possibility of doing is based on the following.

Lemma 4.25 *Let* $\mathrm{Trans}(M)$ *be the formula expressing 'M is transitive'. Let* $\phi(x)$ *be a formula with one free variable x and suppose that there is a finite conjunction $\tau_1 \wedge \cdots \wedge \tau_n = \tau$ of axioms of ZFC such that*

$$\vdash [\mathrm{Trans}(M) \wedge |M| = \aleph_0 \wedge \tau^{(M)}]$$

$$\to [\forall B[B \text{ is a complete Boolean algebra} \to \phi(B)]]^{(M)}. \qquad (1)$$

Then

$$ZFC \vdash \forall B[B \text{ is a complete Boolean algebra} \to \phi(B)].$$

Proof Let σ be the sentence

$$\forall B[B \text{ is a complete Boolean algebra} \to \phi(B)].$$

Then, by the reflection principle and the downward Löwenheim–Skolem theorem, we have

$$ZFC \vdash \exists M[\mathrm{Trans}(M) \wedge |M| = \aleph_0 \wedge \tau^{(M)} \wedge (\sigma^{(M)} \leftrightarrow \sigma)].$$

The result now follows immediately from (1). $\qquad\square$

This lemma enables us to 'transfer' to V results about $V^{(B)}$ derived in M (i.e. results about $M^{(B)}$) in the following way. Suppose that $\Phi(V^{(B)})$ is some first-order statement about $V^{(B)}$, which is expressible as a formula $\phi(B)$. Then, if we can show that $\Phi(V^{(B)})$ holds in each countable transitive model of some finite conjunction of axioms of ZFC, it will follow from the lemma that $\Phi(V^{(B)})$ is a theorem of ZFC, that is, $\Phi(V^{(B)})$ 'holds in V'.

This technique can be applied to elucidate the nature of the canonical generic ultrafilter U_*. First, we note that the same definition of U_* with M replaced by V makes U_* a member of $V^{(B)}$; so we may refer to U_* also as the *canonical generic ultrafilter* in $V^{(B)}$.

Next, we introduce a new constant symbol $\hat{V}$ into $\mathcal{L}^{(B)}$, it being understood that $\hat{V}$ represents a *class*. We extend the assignment of Boolean values to sentences of this augmented language by putting, for $x \in V^{(B)}$,

$$[\![x \in \hat{V}]\!] = \bigvee_{y \in V} [\![x = \hat{y}]\!].$$

Thus $\hat{V}$ represents the class of all *standard* objects in $V^{(B)}$. One can now show that

$$V^{(B)} \models \hat{V} \text{ *is a transitive model of* ZFC *containing all the ordinals.*}$$

Moreover

$$V^{(B)} \models (\hat{B} \text{ *is a complete Boolean algebra* })^{(\hat{V})}.$$

So, working inside $V^{(B)}$, we can construct the $\hat{B}$-extension $\hat{V}^{(\hat{B})}$ of $\hat{V}$. Upon interpreting Theorem 4.21 in $V^{(B)}$, with $\hat{V}$ playing the role of M, we see that

$$V^{(B)} \models U_* \text{ *is a* } \hat{V}\text{—*generic ultrafilter in* } \hat{B}.$$

Accordingly, in $V^{(B)}$ we can form the quotient $\hat{V}^{(\hat{B})}/U_*$ and its transitive collapse $\hat{V}[U_*]$.

Applying Theorem 4.22 within $V^{(B)}$ (with $\hat{V}$ playing the role of M and $\hat{B}$ that of B) we have

$$V^{(B)} \models \hat{V}[U_*] \text{ *is the model of* } \text{ZFC } \text{*generated by* } U_* \text{ *and* } \hat{V}.$$

Also, using Theorem 4.22 and Problem 4.26 one can show that, for any countable transitive $\in$-model M of ZFC, the statement

$$V^{(B)} \models \forall x(x \in \hat{V}[U_*]) \tag{$*$}$$

holds in M. Moreover, in order to derive $(*)$ in M, one only requires the conjunction of a finite number of axioms of ZFC to hold there. Hence, by the remarks following Lemma 4.25, $(*)$ *holds in the real world* (i.e. V) for every complete

Boolean algebra B. Therefore, if we identify $\hat{V}$ with V, we may regard $V^{(B)}$ as the Boolean-valued model of ZFC generated by U_* and V, or as the Boolean-valued model obtained by *adjoining* the B-valued set U_* to V. It is precisely for this reason that we call $V^{(B)}$ a *Boolean extension* of V.

Problems

Throughout, M is a transitive $\in$-model of ZFC and B is a complete Boolean algebra in the sense of M.

4.26 (Truth in $M^{(B)}$) Suppose that M is countable, let $\phi(v_1, \ldots, v_n)$ be a formula and let $x_1, \ldots, x_n \in M^{(B)}$. Show that $M^{(B)} \models \phi(x_1, \ldots, x_n)$ iff $M[U] \models \phi[i_U(x_1), \ldots, i_U(x_n)]$ for *every* M-generic ultrafilter U, where i_U is the canonical map of $M^{(B)}$ onto $M[U]$. (Use Theorem 4.23 and Lemma 4.11.) Hence obtain a new proof of Theorem 4.21. (Use Lemma 4.25.)

4.27 (Countably M-complete ultrafilters) An ultrafilter U in B is said to be *countably M-complete* if whenever $X \in P^{(M)}(B)$ is *countable* in M, we have

$$\bigvee X \in U \leftrightarrow X \cap U \neq \emptyset.$$

Put $N = M^{(B)}/U$. Show that the following are equivalent:

(i) U is countably M-complete;

(ii) $\omega^{(N)}$ is well-ordered under $\in_U$;

(iii) $\omega^{(N)} = \{\hat{n}^U : n \in \omega\}$.

(For (i) $\rightarrow$ (iii), argue as in Theorem 4.6. For (ii) $\rightarrow$ (i), assume (i) fails, choose a partition of unity $\{a_m : m \in \omega\} \in M$ with $a_m \notin U$ for all $m \in \omega$; using the Mixing Lemma define $s_n = \Sigma_{m>n} a_m \cdot (m-n)\hat{\ }$ for each $n \in \omega$. Now show that $\{s_n^U : n \in \omega\}$ is a descending sequence of members of $\omega^{(N)}$.)

4.28 (Atoms in B) Let a be an *atom* in B, *i.e.* such that $a \neq 0$ and, for any $x \in B, x \leq a \rightarrow x = 0$ or $x = a$.

(i) Show that the set $U_a = \{x \in B : a \leq x\}$ is an ultrafilter in B (called the ultrafilter *generated* by a).

(ii) Show that U_a is M-generic, and that $U_a \in M$. Deduce that $M[U_a] = M$.

(iii) Let U be an ultrafilter in B such that $U \in M$, and put $a = \bigwedge U$. Show that the following are equivalent: (a) $a \neq 0$, (b) a is an atom, (c) $U = U_a$.

4.29 (Atoms and $M[U]$) Let A be the set of all atoms in B, and let U_* be the canonical generic ultrafilter in $M^{(B)}$.

(i) Show that $\bigvee_{y \in M} \llbracket U_* = \hat{y} \rrbracket = \bigvee A$. (Observe that $\llbracket U_* = \hat{y} \rrbracket \leq \llbracket \hat{y}$ *is a* $(P^{(M)}(B))\hat{\ }$-*complete ultrafilter in* $\hat{B} \rrbracket = 0$ or 1, and use Problem 4.28 (iii).)

(ii) Show that $\bigwedge_{x\in M^{(B)}} \bigvee_{y\in M} [\![x = \hat{y}]\!] = \bigvee A$. (For any atom $a \in B$, and any $x \in M^{(B)}$, show, using Problem 4.28(i) and Lemma 4.11 that $a \leq \bigvee_{y\in M} [\![x = \hat{y}]\!]$.)

(iii) Assume that $M \models V = L$. Show that

$$[\![V = L]\!] = \bigwedge_{x\in M^{(B)}} \bigvee_{y\in M} [\![x = \hat{y}]\!] = \bigvee_{y\in M} [\![U_* = \hat{y}]\!] = [\![L(U_*)]\!].$$

(Use Theorem 1.46 and (i).)

(iv) Put $\eta = 1$ if $M \models V = L$ and $\eta = 0$ if $M \models V \neq L$. Show that $[\![V = L]\!] = \eta \wedge \bigvee A$.

(v) Assume that $M \models V = L$, and let U be an M-generic ultrafilter in B. Show that $M[U] \models V = L$ iff $U = U_a$ for some atom $a \in B$. (Use (iv).)

4.30 (A trivial Boolean extension) Let $u \in M$, and put $B = P^{(M)}(u)$. B is then the power set Boolean algebra of u in M.

(i) Show that, for any formula $\phi(v_1, \ldots, v_n)$, and any $x_1, \ldots, x_n \in M$,

$$M \models \phi[x_1, \ldots, x_n] \leftrightarrow M^{(B)} \models \phi(\hat{x}_1, \ldots, \hat{x}_n).$$

(If $[\![\phi(\hat{x}_1, \ldots, \hat{x}_n)]\!] \neq 1$, let a be an atom $\leq [\![\neg\phi(\hat{x}_1, \ldots, \hat{x}_n)]\!]$ and use Problem 4.28 (ii).)

(ii) Let U be any ultrafilter in B. Show that, for any sentence σ,

$$M \models \sigma \leftrightarrow M^{(B)}/U \models \sigma.$$

4.31 (A transitive model of $\neg$AC) Let $G \in M$ be a group acting on B, and let $\Gamma \in M$ be a filter of subgroups of G. Put $M^{(\Gamma)} = (V^{(\Gamma)})^{(M)}$ (for the definition of $V^{(\Gamma)}$, see Chapter 3). Let U be an M-generic ultrafilter in B. Recalling that i is the natural map of $M^{(B)}$, onto $M[U]$, put $M[\Gamma, U] = \langle i[M^{(\Gamma)}], \in |i[M^{(\Gamma)}]\rangle$.

(i) Show that $M \subseteq M[\Gamma, U]$, and that $M[\Gamma, U]$ is transitive.

(ii) Show that, for any formula $\phi(v_1, \ldots, v_n)$, and any $x_1, \ldots, x_n \in M^{(\Gamma)}$,

$$M[\Gamma, U] \models \phi[i(x_1), \ldots, i(x_n)] \leftrightarrow [\![\phi(x_1, \ldots, x_n)]\!]^{\Gamma} \in U.$$

(iii) Show that, if M is countable, then for a suitable choice of B, Γ and U, $M[\Gamma, U]$ is a countable transitive model of ZF in which AC fails. (Use Theorem 3.21)

4.32 (The converse to Corollary 4.7 fails)[2] Suppose that there is a measurable cardinal $\mu > \omega$ and an inaccessible cardinal $\kappa > \mu$. Then it follows

---

[2]This problem assumes an acquaintance with measurable cardinals; *cf.* Drake (1974).

that $M = \langle R_\kappa, \in | R_\kappa \rangle$ is a transitive $\in$-model of ZFC. Let U be a μ-complete nonprincipal ultrafilter in $P\mu \in M$. Show that $M^{(P\mu)}/U$ is well-founded but U is not M-generic. (Note that, by Problem 3.13, $M^{(P\mu)}/U$ is isomorphic to the ultrapower M^μ/U.)

4.33 (Construction of uncountable transitive models of ZFC+$V \neq L$)

(i) Let κ be an infinite cardinal, and suppose that the complete Boolean algebra B contains a κ-closed dense subset P (Problem 2.17). Show that, for any $S \subseteq PB$ such that $|S| \leq \kappa$, there is an S-complete ultrafilter in B. (Let $S = \{T_\xi : \xi < \kappa\}$ and first confine attention to the case in which $\bigvee T_\xi = 1$ for all $\xi < \kappa$. Let J be a sufficiently large index set so that each T_α can be enumerated as $\{t_{\xi j} : j \in J\}$. Using the fact that P is κ-closed, construct by transfinite recursion a function $f \in J^\kappa$ such that $\bigwedge_{\xi < \alpha} t_{\xi f(\xi)} \neq 0$ for each $\alpha < \kappa$. Conclude that there is an ultrafilter which intersects each T_ξ; now apply this to the general case.)

(ii) Let κ be a regular cardinal. Show that, for each family S of subsets of $B_\kappa(\kappa, 2)$ (Problem 2.18) such that $|S| \leq \kappa$, there is an S-complete ultrafilter in $B_\kappa(\kappa, 2)$. (Use (i) and Problem 2.18(ii).)

(iii) Suppose that there exists an inaccessible cardinal $\lambda > \omega$. Show that, for each infinite cardinal $\kappa < \lambda$ there is a transitive $\in$-model of ZFC $+ V \neq L$ of cardinality κ. (By Löwenheim–Skolem it is enough to prove the result for all *regular* $\kappa < \lambda$. So let κ be regular, put $B = B_\kappa(\kappa, 2)$ and $M = \langle R_\lambda, \in | R_\lambda \rangle$. Using the Maximum Principle in $M^{(B)}$, for each formula $\phi(v_0, \ldots, v_n)$ let $f_\phi : (M^{(B)})^n \to M^{(B)}$ be a 'Skolem function for ϕ in $M^{(B)}$', that is, such that, for all $x_1, \ldots, x_n \in M^{(B)}$, $[\![\exists v_0 \phi(v_0, x_1, \ldots, x_n)]\!] = [\![\phi(f_\phi(x_1, \ldots, x_n), x_1, \ldots, x_n)]\!]$. Let A be the closure of the set $\{\hat{\xi} : \xi < \kappa\}$ under all the f_ϕ. By (ii), let U be an S-complete ultrafilter in B, where $S = \{\{[\![a = \hat{\xi}]\!] : \xi < \kappa\} : a \in A\}$. Show that the structure $\langle \{a^U : u \in A\}, \in_U \rangle$ is a well-founded model of ZFC+$V \neq L$ of cardinality κ.)

4.34 (Generic sets of conditions) Let $\langle P, \leq \rangle$ be a partially ordered set in M. A subset X of P is said to be *dense* in P if $\forall y \in P \exists x \in X[x \leq y]$, and *dense below an element* $p \in P$ if $\forall y \leq p \exists x \in X[x \leq y]$. A subset G of P is said to be *M-generic* if

(a) $x \in G, y \in P, x \leq y \to y \in G$;

(b) $\forall xy \in G \exists z \in G[z \leq x \wedge z \leq y]$;

(c) $G \cap X \neq \emptyset$ for every dense subset X of P, which is in M.

(i) Show that, if G is M-generic in P and $X \subseteq P$ is dense below an element of G, then $X \cap G \neq \emptyset$.

Now let Q be the refined associate of P (Problem 2.4) and let $j : P \to Q$ be the canonical map (if P is refined then $P = Q$ and j is the identity).

Let $B = \mathrm{RO}(Q)^{(M)}$ be the Boolean completion of Q in M, and identify Q as a dense subset of B.

(ii) Show that, if G is M-generic in P, then

$$\overline{G} = \{x \in B : \exists y \in G[j(y) \le x]\}$$

is an M-generic ultrafilter (called the M-generic ultrafilter *generated* by G). Show also that $G = j^{-1}[\overline{G}]$. Show conversely, that, if U is an M-generic ultrafilter in B, then $j^{-1}[U]$ is an M-generic subset of P. Deduce that, if M is *countable*, P has an M-generic subset.

It follows from (ii) that, if U is the M-generic ultrafilter generated by an M-generic subset G of P, then G is definable from U and vice-versa. *Under these circumstances we write $M[G]$ for $M[U]$.* If $p \in P$, and σ is an $\mathcal{L}_M^{(B)}$-sentence, we write $p \Vdash \sigma$ for $j(p) \Vdash \sigma$.

(iii) Show that, if G is M-generic in P, $M[G]$ is the least transitive model of ZF, which includes M and contains G.

(iv) Let G be M-generic in P and let i be the canonical map of $M^{(B)}$ onto $M[G]$. Show that, for any formula $\phi(v_1, \ldots, v_n)$ and any $x_1, \ldots, x_n \in M^{(B)}$,

$$M[G] \models \phi[i(x_1), \ldots, i(x_n)] \leftrightarrow \exists p \in G[p \Vdash \phi(x_1, \ldots, x_n)].$$

(v) *Iteration lemma.* Let P be a refined partially ordered set in M, let G be an M-generic subset of P, let Q be a refined partially ordered set in $M[G]$, and let H be an $M[G]$-generic subset of Q. Show that there is a refined partially ordered set R in M and an M-generic subset K of R such that $M[G][H] = M[K]$. (Let $\le_Q$ be the partial ordering of Q. First show that without loss of generality we may assume that Q (but not $\le_Q$) is in M. Let B be the Boolean completion of P in M, let i be the canonical map of $M^{(B)}$ onto $M[G]$, let $\le^*$ be an element of $M^{(B)}$ such that $i(\le^*) = \le_Q$, and let σ be the $\mathcal{L}_M^{(B)}$-sentence: $\langle \hat{Q}, \le^* \rangle$ *is a refined partially ordered set*. Observe that there is $p \in G$ for which $p \Vdash \sigma$. Now let $P' = \{p \in P : p \Vdash \sigma\}$, $R = P' \times Q$ and define the relation $\le$ on R by $\langle p_1, q_1 \rangle \le \langle p_2, q_2 \rangle$ iff ($p_1 \le p_2$ and $p_1 \Vdash \hat{q}_1 \le^* \hat{q}_2$). Show that $\langle R, \le \rangle$ is a refined partially ordered set in M. Now put $G' = G \cap P'$ and $K = G' \times H$. Show that every dense subset of P' meets G'; use this, together with the genericity of G and H to prove that K is M-generic. Finally, use the fact that $G = \{p \in P : \exists r \in G'[r \le p]\}$ to prove the last assertion.)

(vi) *Product lemma.* Let P and Q be refined partially ordered sets in M, let G be an M-generic subset of P and let H be an $M[G]$-generic subset of Q. Give $P \times Q$ the product ordering: $\langle p_1, q_1 \rangle \le \langle p_2, q_2 \rangle$ iff $p_1 \le p_2$ and $q_1 \le q_2$. Show that $G \times H$ is an M-generic subset of $P \times Q$ and that $M[G \times H] = M[G][H]$. (Like (v).)

4.35 (Canonical generic sets and the adjunction of maps) Let P be a basis for B in M. Define $G_* \in M^{(B)}$ by $\mathrm{dom}(G_*) = \{\hat{p} : p \in P\}, G_*(\hat{p}) = p$ for $p \in P$.

(i) Let G be M-generic in P, and let $i : M^{(B)} \to M[G]$ be the canonical map. Show that $i(G_*) = G$. (Like Theorem 4.19.)

(ii) Show that

$$M^{(B)} \models G_* \text{ is a generic subset of } \hat{P}.$$

(Like Theorem 4.21.) For this reason G_* is called the *canonical generic set* in $M^{(B)}$.

Now define $G_{**} \in M^{(B)}$ by $\mathrm{dom}(G_{**}) = \bigcup\{\mathrm{dom}(y) : y \in \mathrm{dom}(G_*)\}$ and $G_{**}(x) = [\![\exists y \in G_*[x \in y]]\!]$ for $x \in \mathrm{dom}(G_{**})$.

(iii) Show that $M^{(B)} \models G_{**} = \bigcup G_*$

Now suppose that a, b are nonempty elements of M such that $|b| \geq 2$ and $\aleph_0 \leq |a|$ in M. Let G be an M-generic subset of $P = C(a,b)^{(M)}$ and let $B = \mathrm{RO}(b^a)^{(M)}$.

(iv) Show that $M^{(B)} \models G_{**}$ *is a map of $\hat{a}$ onto $\hat{b}$.* (See the proof of Theorem 5.1.)

(v) Show that $M[G] \models \bigcup G$ *is a map of a onto b.*

Results (iv) and (v) show that, for this choice of B and G, in $M^{(B)}$ we have adjoined the canonical 'map' G_{**} of $\hat{a}$ onto $\hat{b}$ and in $M[G]$ we have adjoined the map $\bigcup G$ of a onto b. Notice that if a is (really) *countable* and b is (really) *uncountable*, no transitive model of ZF can contain a map of a onto b. It follows that, if M is uncountable, there may be no M-generic subset of $C(a,b)^{(M)}$ and hence no M-generic ultrafilters in $\mathrm{RO}(b^a)^{(M)}$. On the other hand,

(vi) If M is *countable*, show that there is an M-generic subset of P. (Use Theorem 4.23 and Problem 4.34(ii).)

(vii) Let M be countable, put $P = C(\omega, 2)^{(M)}$ and let G be an M-generic subset of P. Show that, in $M[G], \bigcup G$ is a nonconstructible map of ω into 2. (Like Theorem 2.6.)

(viii) Let M be countable, and suppose that $M \models GCH$. Put $P = C(\omega \times \omega_2, 2)^{(M)}$ and let G be an M-generic subset of P. Show that, in $M[G], \bigcup G$ is a map of $\omega \times \omega_2^{(M)}$ onto 2. For each $\nu < \omega_2^{(M)}$, put $u_\nu = \{n \in \omega : (\bigcup G)(n, \omega) = 1\}$. Show that $\{u_\nu : \nu < \omega_2^{(M)}\}$ is a set of $\aleph_2^{(M)} = \aleph_2^{(M[G])}$ subsets of ω in $M[G]$. (Like Theorem 2.12.)

4.36 (Adjunction of a subset of ω) Let P be a basis for B in M, and let G be an M-generic subset of P. If $s \in M[G], s \subseteq m$ (or equivalently, if $\hat{s} \in M[U]$, $s \subseteq M$ where U is the M-generic ultrafilter generated by G, *cf.* Problem 4.34), it can be shown—although we do not prove it here (for a proof see Grigorieff 1975)—that there is a least transitive model $M[s]$ of ZF, which includes $M \cup \{s\}$. $M[s]$ is called the model of ZF obtained by *adjoining* s to M: it has the following basic properties:

(1) $M[s] \subseteq M[G]$;

(2) $M[s] \models AC$;

(3) if $t \in M[G]$ is absolutely definable from s and elements of M, then $t \in M[s]$.

> (i) Let $a, b \in M$; let G be an M-generic subset of $C(a,b)^{(M)}$ and let $F = \bigcup G$. Show that $M[F] = M[G]$. (Note that we have $G = \{f \in C(a,b)^{(M)} : f \subseteq F\}$.)
>
> (ii) Let κ be an infinite cardinal in M; let G be a generic subset of $C(\omega, \kappa)^{(M)}$, and let $F = \bigcup G$. Show that there is $s \subseteq \omega, s \in M[F]$ such that $M[F] = M[s]$. (Put $s = \{2^n 3^m : F(n) \leq F(m)\} \in M[F]$. Put $m \sim n$ iff $F(m) = F(n)$; the set A of equivalence classes $\tilde{m}$ can be ordered by $\tilde{m} < \tilde{n}$ iff $2^n 3^m \notin s$. Show that F induces an order preserving map $F' : \langle A, < \rangle \to \langle \kappa, < \rangle$, so that $\langle A, < \rangle$ is well-ordered in $M[s]$. Let $F'' : A \to \alpha$ be an isomorphism of A with an ordinal in $M[s]$. Show that $F'' \circ (F')^{-1}$ is the identity, and conclude that $F \in M[s]$.)

4.37 (Intermediate submodels and complete subalgebras) Let $X \in P^{(M)}(B)$. The complete subalgebra (in M) of B *generated* by X is defined to be the least complete subalgebra of B in M which includes X.

> (i) Suppose that $|B| = \kappa$ (in M), and let B' be the complete subalgebra of B generated by X. Define the sets $\{B_\alpha : \alpha < \kappa^{+(M)}\}$ inductively as follows: $B_0 = X$; if α is odd, $B_\alpha = \{b^* : b \in \bigcup_{\beta < \alpha} B_\beta\}$; if α is even, $B_\alpha = \{\bigvee Y : Y \in M \text{ and } Y \subseteq \bigcup_{\beta < \alpha} B_\beta\}$; if α is a limit, $B_\alpha = \bigcup_{\beta < \alpha} B_\beta$. Show that
> $$B' = \bigcup_{\alpha < \kappa^{+(M)}} B_\alpha.$$

From now on we let U be an M-generic ultrafilter in B, let $i : M^{(B)} \to M[U]$ be the canonical map.

> (ii) Let $s \in M[U]$ and $s \subseteq M$. Show that there is $t \in M$ such that $s \subseteq t$, and hence $s_* \in M^{(B)}$ such that $i(s_*) = s$ and $\mathrm{dom}(s_*) = \hat{t}$. (For each $x \in M[U]$ let $\underline{x} \in M^{(B)}$ be such that $i(\underline{x}) = x$. Then $s \subseteq M$ means that $\bigvee_{y \in M} [\![\underline{x} = \hat{y}]\!] \in U$ for each $x \in s$. Now argue as in the proof of Lemma 1.36. For the second assertion, consult the proof of Lemma 1.38.)
>
> (iii) Let $s \in M[U]$ and $s \subseteq M$. Let $B(s_*)$ be the complete subalgebra of B generated by $\{[\![\hat{x} \in s_*]\!] : x \in M\}$, where s_* is as in (ii). Show that $M[s] = M[U \cap B(s_*)]$. (Notice first that the equation in question makes sense because $U \cap B(s_*)$ is an M-generic ultrafilter in $B(s_*)$ and $s_* \in M^{(B(s_*))}$. Next, show that $s \in M[U \cap B(s_*)]$, so that $M[s] \subseteq M[U \cap B(s_*)]$. Now prove the reverse inclusion by showing that, if N is any transitive model of

ZF and $M \cup \{s\} \subseteq N$, then $U \cap B(s_*) \in N$. Put $U_\alpha = U \cap B_\alpha$, where the B_α are defined as in (i), with $X = \{[\![\hat{x} \in s_*]\!] : x \in M\}$. Show by induction that $U_\alpha \in N$ for all $\alpha < (|B|^+)^{(M)}$. Conclude that $U \cap B(s_*) = \bigcup_\alpha B_\alpha \in N$.)

(iv) Let N be a transitive model of ZFC such that $M \subseteq N \subseteq M[U]$. Show that $N = \bigcup \{M[s] : s \subseteq M \text{ and } s \in N\}$. (Given $x \in N$, show that there is $s \in N$ with $s \subseteq \mathrm{ORD}^{(N)} = \mathrm{ORD}^{(M)}$ such that $x \in M[s]$ as follows. Since AC holds in N, there is a bijection f in N of a cardinal $\kappa \in N$ onto the transitive closure t of x. Define $r \subseteq \kappa \times \kappa$ by $\langle \alpha, \beta \rangle \in r \leftrightarrow f(\alpha) \in f(\beta)$. Let g be the canonical map of $\kappa \times \kappa$ onto κ (in M), and put $s = g[r]$.)

(v) Let N be a transitive model of ZFC such that $M \subseteq N \subseteq M[U]$. Show that there is a complete subalgebra A of B such that $N = M[U \cap A]$. (By (iv), choose $s \subseteq M, s \in N$ such that $P^{(N)}(B) \in M[s] \subseteq N$. Now use (iii) and (iv) to get $N = M[s]$ and apply (iii) again.)

(vi) B is said to be *countably generated* (in M) if there is a countable subset X of B in M such that the complete subalgebra of B generated by X is B itself. Show that, if B is countably generated, then there is an $s \in M[U]$, $s \subseteq \omega$ such that $M[U] = M[s]$. (Let $X = \{b_n : n \in \omega\}$ be the countable generating set. Define $s' \in M^{(B)}$ by $\mathrm{dom}(s') = \hat{\omega}$ and $s'(\hat{n}) = b_n$. Use (iii) to show that $s = i(s')$ meets the requirements.)

4.38 (Involutions and generic ultrafilters) Let U be an M-generic ultrafilter in B and let $\pi \in M$ be an automorphism of B.

(i) Show that $\pi[U]$ is an M-generic ultrafilter in B and that $M[U] = M[\pi[U]]$.

(ii) Let $f : B \to B, f \in M$ be such that $f[U] \subseteq U$. Show that there is a $b \in U$ such that $f(x) \geq x$ for all $x \leq b$. (Put $b = \bigwedge \{x^* \vee f(x) : x \in B\}$.)

(iii) An M-*involution* of B is an automorphism $\pi \in M$ of B such that π^2 is the identity. Let $W \subseteq B, W \in M[U]$. Show that the following conditions are equivalent:

(a) W is an M-generic ultrafilter in B and $M[U] = M[W]$;

(b) there is an M-involution π of B such that $\pi[U] = W$. (For (b) $\to$ (a), use (i). For (a) $\to$ (b), assume (a) and $U \neq W$. Let i_U, i_W be the canonical maps of $M^{(B)}$ onto $M[U], M[W]$ respectively and let $\underline{U}, \underline{W}$ be such that $i_w(\underline{U}) = U, i_U(\underline{W}) = W$. Define the functions k, l, m, n of B into B by $k(x) = [\![\hat{x} \in \underline{U}]\!], m(x) = [\![\hat{x} \in \underline{W}]\!], l(x) = \bigwedge \{y \in B : x \leq m(y)\}, n(x) = \bigwedge \{y \in B : x \leq k(y)\}$. Now put $f(x) = k(x) \wedge l(x), g(x) = m(x) \wedge n(x)$. Show that $(g \circ f)[U] \subseteq U$ and $(f \circ g)[W] \subseteq W$ and that $(g \circ f)(x) \leq x, (f \circ g)(x) \leq x$. By (ii), choose $b_0 \in U, c_0 \in W$ such that $(g \circ f)(x) = x$ for all $x \leq b_0$ and $(f \circ g)(x) = x$ for all $x \leq c_0$. Since $U \neq W$, we may assume that $b_0 \wedge c_0 = 0$. Put $b = b_0 \wedge g(c_0) \in U$ and $c = f(b) \in W$, and let $B_b = \{x \in B : x \leq b\}, B_c = \{x \in B : x \leq c\}$. Show that $f|B_b$ is an isomorphism of B_b onto B_c and $g|B_c$ is the inverse of $f|B_b$. Now put, for $x \in B, \pi(x) = f(x \wedge b) \vee g(x \wedge c) \vee (x - (b \wedge c))$.)

4.39 (The submodel of hereditarily ordinal definable sets) Let $\mathcal{L}_S$ be the extension of the language $\mathcal{L}$ of set theory to include a new unary predicate symbol S, and let $\mathcal{L}_S^{(B)}$ be the language obtained for $\mathcal{L}_M^{(B)}$ by adding S. We extend the assignment of Boolean values to $\mathcal{L}_S^{(B)}$-sentences by defining, for $x \in M^{(B)}$, $[\![S(x)]\!] = \bigvee_{y \in M}[\![x = \hat{y}]\!]$. Let U be an M-generic ultrafilter in B, and let i_U be the canonical map of $M^{(B)}$ onto $M[U]$. We write $(M[U], M)$ for the $\mathcal{L}_S$-structure obtained from $M[U]$ by interpreting S as the subset M of $M[U]$.

(i) Show that for any $\mathcal{L}_S$-formula $\phi(v_1, \ldots, v_n)$ and any $x_1, \ldots, x_n \in M^{(B)}$

$$(M[U], M) \models \phi[i_U(x_i), \ldots, i_U(x_n)] \leftrightarrow [\![\phi(x_1, \ldots, x_n)]\!] \in U.$$

Now put $(\text{ODM})^{M[U]}$ for the collection of all elements of $M[U]$ which are definable in $(M[U], M)$ from (ordinals and) elements of M, $(\text{HODM})^{M[U]}$ for the collection of all elements of $(\text{ODM})^{M[U]}$ whose transitive closure is in $(\text{ODM})^{M[U]}$. $(\text{ODM})^{M[U]}$ and $(\text{HODM})^{M[U]}$ are the sets of elements of $M[U]$ which are ordinal definable, and hereditarily ordinal definable, respectively, from M in $M[U]$. It can be shown that $(\text{HODM})^{M[U]}$ is a transitive model of ZFC; evidently $M \subseteq (\text{HODM})^{M[U]} \subseteq M[U]$, so by Problem 4.37(v) there is a complete subalgebra A of B such that $(\text{HODM})^{M[U]} = M[U \cap A]$. We now describe A explicitly.

Let $B^+ = \{x \in B : \pi(x) = x \ for \ every \ automorphism \ \pi \in M \ of \ B\}$. It is easy to see that B^+ is a complete subalgebra of B.

(ii) Show that $(\text{HODM})^{M[U]} = M[U \cap B^+]$. (Put $X = \{W \in M[U] : W \ is \ an \ M\text{-}generic \ ultrafilter \ in \ B \ and \ M[U] = M[W]\}$. By Problem 4.38 $X = \{\pi[U] : \pi \in M \ is \ an \ automorphism \ of \ B\}$. Hence show that $U \cap B^+ = \bigcap\{W \cap B^+ : W \in X\}$ and that $U \cap B^+$ is definable in $(M[U], M)$ from B. Infer that $U \cap B^+ \in (\text{HODM})^{M[U]}$, so that $M[U \cap B^+] \subseteq (\text{HODM})^{M[U]}$. To establish the reverse inclusion, obeserve that it suffices to show that, if $s \in (\text{HODM})^{M[U]}$ and $s \subseteq M[U \cap B^+]$, then $s \in M[U \cap B^+]$ (for if $(\text{HODM})^{M[U]} - M[U \cap B^+] \neq \emptyset$, an element of it of minimal rank would be a subset of $M[U \cap B^+]$). Since $U \cap B^+$ is definable in $(M[U], M)$ from B, so is the canonical map of $M^{(B^+)}$ onto $M[U \cap B^+]$, so we may suppose that $s \subseteq M$. Let $t \in M$ be such that $s \subseteq t$, let $x_0, \ldots, x_n \in M$ and let $\phi(v_0, \ldots, v_{n+1})$ be an $\mathcal{L}_s$-formula such that for all $x \in t$,

$$x \in s \leftrightarrow (M[U], M) \models \phi(x, x_0, \ldots, x_n).$$

Notice that $[\![\phi(\hat{x}, \hat{x}_0, \ldots, \hat{x}_n)]\!] \in B^+$, and that $x \in s$ if and only if $[\![\phi(\hat{x}, \hat{x}_0, \ldots, \hat{x}_n)]\!] \in U \cap B^+$. Conclude that $s \in M[U \cap B^+]$.)

5

CARDINAL COLLAPSING, BOOLEAN ISOMORPHISM, AND APPLICATIONS TO THE THEORY OF BOOLEAN ALGEBRAS

Cardinal collapsing

We have seen in Chapter 1 that if the complete Boolean algebra B satisfies the countable chain condition, then cardinals in V retain their true size in $V^{(B)}$. In this section we show that, if B does not satisfy this condition, it becomes possible for two infinite cardinals $\kappa < \lambda$ to satisfy $V^{(B)} \models |\hat{\lambda}| = |\hat{\kappa}|$. In this event we say that λ has been *collapsed* to κ in $V^{(B)}$. We begin by formulating a necessary and sufficient condition on B for this to happen.

Theorem 5.1 *Let κ and λ be infinite cardinals with $\kappa \leq \lambda$. Then the following conditions are equivalent:*

(i) $V^{(B)} \models |\hat{\kappa}| = |\hat{\lambda}|$;

(ii) *there is a double sequence $\{b_{\xi\eta} : \xi < \kappa, \eta < \lambda\} \subseteq B$ such that $\bigvee_{\xi<\kappa} b_{\xi\eta} = 1$ for all $\eta < \lambda$ and $\{b_{\xi\eta} : \eta < \lambda\}$ is an antichain for each $\xi < \kappa$.*

Proof (i) $\rightarrow$ (ii). Suppose (i) holds. Then we have

$$V^{(B)} \models \exists f[f \text{ is a map of } \hat{\kappa} \text{ onto } \hat{\lambda}].$$

Using the Maximum Principle, it follows that there is $f \in V^{(B)}$ such that

$$V^{(B)} \models f \text{ is a map of } \hat{\kappa} \text{ onto } \hat{\lambda}. \tag{1}$$

Put $b_{\xi\eta} = [\![f(\hat{\xi}) = \hat{\eta}]\!]$ for $\xi < \kappa, \eta < \lambda$. Then if $\eta, \eta' < \lambda$ and $\eta \neq \eta'$,

$$b_{\xi\eta} \wedge b_{\xi\eta'} = [\![f(\hat{\xi}) = \hat{\eta} \wedge f(\hat{\xi}) = \hat{\eta}']\!] \leq [\![\hat{\eta} = \hat{\eta}']\!] = 0,$$

and, for $\eta < \lambda$,

$$\bigvee_{\xi<\kappa} b_{\xi\eta} = \bigvee_{\xi<\kappa} [\![f(\hat{\xi}) = \hat{\eta}]\!] = [\![\exists x \in \hat{\kappa}[f(x) = \hat{\eta}]\,]\!] = 1,$$

by (1). Thus $\{b_{\xi\eta} : \xi < \kappa, \eta < \lambda\}$ satisfies (ii).

(ii) $\rightarrow$ (i). Assume (ii). Since $\kappa < \lambda$, we have $V^{(B)} \models |\hat{\kappa}| \leq |\hat{\lambda}|$, so it suffices to show that $V^{(B)} \models |\hat{\lambda}| \leq |\hat{\kappa}|$. To this end, define $f \in V^{(B)}$ by

$$\mathrm{dom}(f) = \{\langle \hat{\xi}, \hat{\eta} \rangle\}^{(B)} : \xi < \kappa, \eta < \lambda\}$$

and, for $\xi < \kappa, \eta < \lambda$,

$$f(\langle \hat{\xi}, \hat{\eta} \rangle^{(B)}) = b_{\xi\eta}.$$

Using the assumption that $\{b_{\xi\eta} : \eta < \lambda\}$ is an antichain for each $\xi < \kappa$, it follows easily that

$$V^{(B)} \models f \text{ is a map with } \mathrm{dom}(f) \subseteq \hat{\kappa} \text{ and } \mathrm{ran}(f) \subseteq \hat{\lambda}.$$

Also, for each $\eta < \lambda$ we have, by assumption,

$$[\![\exists x \in \hat{\kappa}[f(x) = \hat{\eta}]]\!] = \bigvee_{\xi < \kappa} [\![f(\hat{\xi}) = \hat{\eta}]\!] = \bigvee_{\xi < \kappa} b_{\xi\eta} = 1.$$

It follows that $V^{(B)} \models \hat{\lambda} \subseteq \mathrm{ran}(f)$, and so $V^{(B)} \models |\hat{\lambda}| \leq |\hat{\kappa}|$, completing the proof. $\qquad\square$

Let λ be an infinite cardinal, let X be the product space λ^ω, where λ is assigned the discrete topology, and let $B = \mathrm{RO}(X)$. For each $m \in \omega$ and $\eta < \lambda$ let $b_{m\eta} = \{g \in X : g(m) = \eta\}$. It is then straightforward to verify that $\{b_{m\eta} : m \in \omega, \eta < \lambda\}$ is a subset of B satisfying Theorem 5.1(ii), with $\kappa = \omega$. Accordingly, that theorem gives:

Corollary 5.2 *Let $B = \mathrm{RO}(\lambda^\omega)$, where $\lambda \geq \aleph_0$. Then*

$$V^{(B)} \models \hat{\lambda} \text{ is countable.}$$

This result shows that $\mathrm{RO}(\lambda^\omega)$ may be thought of as an algebra which adjoins a collapsing map of $\hat{\omega}$ onto $\hat{\lambda}$; accordingly $\mathrm{RO}(\lambda^\omega)$ is called the *collapsing* $(\aleph_0, \lambda)$-*algebra*. In the next section we shall show that these collapsing algebras have other useful features.

Problems

5.3 ($P\omega \bigcap L$ **can be countable**)

(i) Let $\lambda \geq \aleph_0$ and let B be the collapsing $(\aleph_0, 2^\lambda)$-algebra. Show that

$$V^{(B)} \models P\hat{\lambda} \cap L \text{ is countable.}$$

(ii) Let M be a countable transitive model of ZFC $+\ 2^{\aleph_0} = \aleph_1$, put $B = (\mathrm{RO}(\omega_1^\omega))^{(M)}$, and let U be an M-generic ultrafilter in B. Show that

$$M[U] \models P\omega \cap L \ \text{is countable.}$$

5.4 (More on collapsing algebras) Assume GCH. Let κ, λ be regular infinite cardinals with $\kappa < \lambda$. Put $B = B_\kappa(\kappa, \lambda)$ (*cf.* Problem 2.18).

(i) Show that $V^{(B)} \models \mathrm{Card}(\hat\alpha)$ for any cardinal $\alpha \leq \kappa$. (Use Problems 2.20(i) and 2.18.)

(ii) Show that $V^{(B)} \models \mathrm{Card}(\hat\alpha)$ for any cardinal $\alpha \geq \lambda^+$. (Show that $|C_\kappa(\kappa, \lambda)| = \lambda$ and so B satisfies the $\lambda^+ - $ cc. Now use Problem 1.53.)

(iii) Show that $V^{(B)} \models |\hat\lambda| = \hat\kappa$. (Use Theorem 5.1.)

$B_\kappa(\kappa, \lambda)$ is called the *collapsing* (κ, λ)-*algebra*: its effect is to collapse λ to κ but not to collapse any cardinal $\leq \kappa$ or $\geq \lambda^+$.

5.5 (Consistency of CH and $\neg$CH with the existence of measurable cardinals) Let κ be a cardinal. An ultrafilter F in $P\kappa$ is said to be *nonprincipal* if $\{a\} \notin F$ for all $\alpha < \kappa$, and κ-*complete* if whenever $\alpha < \kappa$ and $\{X_\xi : \xi < \alpha\} \subseteq F$, then $\bigcap_{\xi < \alpha} X_\xi \in F$. The cardinal κ is said to be *measurable* (*cf.* Drake 1974) if $\kappa > \aleph_0$ and there is a nonprincipal κ-complete ultrafilter in $P\kappa$. It is known that, if κ is measurable, then κ is regular and $2^\lambda < \kappa$ for every cardinal $\lambda < \kappa$, that is, κ is inaccessible.

(i) Let κ be a measurable cardinal and let F be a κ-complete nonprincipal ultrafilter in $P\kappa$. Let B be a complete Boolean algebra with a basis P such that $|P| < \kappa$. Define $G \in V^{(B)}$ by $\mathrm{dom}(G) = B^{\mathrm{dom}(\hat\kappa)}$ and, for $y \in \mathrm{dom}(G), G(y) = [\![y \subseteq \hat\kappa \wedge \exists x \in \hat F [x \subseteq y]\,]\!]$.

(a) Show that $V^{(B)} \models \mathrm{Card}(\hat\kappa) \wedge \hat\kappa > \aleph_0$. (Use Problem 1.53(iii).)

(b) Show that $V^{(B)} \models G$ *is a nonprincipal filter in* $P\hat\kappa$. (Use the fact that $[\![u \in G]\!] = [\![u \subseteq \hat\kappa]\!] \wedge \bigvee_{r \in F} [\![\hat x \subseteq u]\!]$.)

(c) Show that $V^{(B)} \models G$ *is an ultrafilter in* $P\hat\kappa$. (For this it suffices to show that, for any $p \in P$ and $u \in V^{(B)}, p \Vdash (u \cap \hat\kappa \in G) \vee (\hat\kappa - u \in G)$. Let $t = \{\alpha < \kappa : p \Vdash \hat\alpha \in u\}$. Show that, if $t \in F$, then $p \models u \cap \hat\kappa \in G$. On the other hand, if $t \notin F$, then $\{\alpha < \kappa : p \nVdash \hat\alpha \in u\} \in F$; using the κ-completeness of F and the fact that $|P| < \kappa$, deduce that there is $q \leq p$ such that $q \Vdash \hat\kappa - u \in G$. Now use Theorem 2.5(iii).)

(d) Show that $V^{(B)} \models G$ *is* $\hat\kappa$-*complete*. (Given $\alpha < \kappa, p \in P$ and $p \Vdash f : \hat\alpha \to G$, it must be shown that $p \Vdash \bigcap_{x \in \hat\alpha} f(x) \in G$. For each $\xi < \alpha$, put $t_\xi = \{\beta < \kappa : p \Vdash \hat\beta \in f(\hat\xi)\}$. Argue as in (c) to derive a contradiction from the assumption that $t_\xi \notin F$. Thus $t_\xi \in F$ for all $\xi < \alpha$; show that $p \Vdash (\bigcap_{\xi < \alpha} t_\xi)\hat{} \subseteq \bigcap_{x \in \hat\alpha} f(x)$, and use the κ-completeness of F.)

(e) Show that $V^{(B)} \models \hat\kappa$ *is a measurable cardinal*. (Use (a)–(d).)

Now let ZFM $=$ ZFC $+$ 'there exists a measurable cardinal'.

(ii) Show that, if ZFM is consistent, so is $\text{ZFM} + 2^{\aleph_0} = \aleph_1$. (Let $P = C_{\omega_1}(\omega_1, P\omega)$ and let $B = B_{\omega_1}, (\omega_1, P\omega)$. Notice that P is $\aleph_1$-closed (Problem 2.17), so that $V^{(B)} \models (P\omega)\hat{} = P\hat{\omega}$. Now use Theorem 5.1 to show that $V^{(B)} \models |(P\omega)\hat{}| \leq \aleph_1$. Conclude that $V^{(B)} \models 2^{\aleph_0} = \aleph_1$, and use (i) (e).)

(iii) Show that, if ZFM is consistent, so is $\text{ZFM} + 2^{\aleph_0} \geq \aleph_2$. (Use (i) (e) and Theorem 2.12.)

Boolean isomorphism and infinitary equivalence

In the previous section we saw that for any pair of infinite sets there is a Boolean extension of V in which they become equipollent. Introducing the idea of an infinitary language, we shall extend this result from sets to structures.

Let $\mathfrak{A} = \langle A, R \rangle$ and $\mathfrak{B} = \langle B, S \rangle$ be structures of the same similarity type, where R and S are binary relations.[1] We write $f : \mathfrak{A} \cong \mathfrak{B}$ if f is an isomorphism of $\mathfrak{A}$ onto $\mathfrak{B}$ and $\mathfrak{A} \cong \mathfrak{B}$ if $f : \mathfrak{A} \cong \mathfrak{B}$ for some f.

Now let C be a complete Boolean algebra. We write $\hat{\mathfrak{A}}$ for the $V^{(C)}$-structure $\langle \hat{A}, \hat{R} \rangle^{(C)}$. We say that $\mathfrak{A}$ and $\mathfrak{B}$ are *C-isomorphic*, written $\mathfrak{A} \cong_C \mathfrak{B}$ if $V^{(C)} \models \hat{\mathfrak{A}} \cong \hat{\mathfrak{B}}$; they are *Boolean isomorphic*, written $\mathfrak{A} \cong_{\mathbf{Bool}} \mathfrak{B}$, if $\mathfrak{A} \cong_C \mathfrak{B}$ for some C.

Next, let $\mathscr{L}$ be the first-order language appropriate for binary structures. The infinitary language $\mathscr{L}_{\infty\omega}$ has the same symbols as $\mathscr{L}$ except that we add a variable x_α for each ordinal α and allow conjunctions and disjunctions of arbitrary sets of formulas in our formation rules; quantifiers, however, are added just one at a time as usual. The class of sentences of $\mathscr{L}_{\infty\omega}$ holding in a structure is called its $\infty\omega$-*theory*. Two structures $\mathfrak{A}$ and $\mathfrak{B}$ with the same $\infty\omega$-theories are said to be $\infty\omega$-*equivalent*; in this event we write $\mathfrak{A} \equiv_{\infty\omega} \mathfrak{B}$.

The relation of $\infty\omega$-equivalence can be characterized in terms of the concept of partial isomorphism. A *partial isomorphism* between $\mathfrak{A}$ and $\mathfrak{B}$ is a nonempty family P of functions such that

- for each $f \in P$, we have $\text{dom}(f) \subseteq A$, $\text{ran}(f) \subseteq B$, and f is an isomorphism of $\mathfrak{A} \restriction \text{dom}(f)$ to $\mathfrak{B} \restriction \text{ran}(f)$;
- if $f \in P, a \in A, b \in B$, then there exist $g, h \in P$, both extending f, such that $a \in \text{dom}(g)$ and $b \in \text{ran}(h)$.

If P is a partial isomorphism between $\mathfrak{A}$ and $\mathfrak{B}$, we write $P : \mathfrak{A} \cong_{\mathbf{part}} \mathfrak{B}$; if there exists such a P we write $\mathfrak{A} \cong_{\mathbf{part}} \mathfrak{B}$ and say that $\mathfrak{A}$ and $\mathfrak{B}$ are *partially isomorphic*.

---

[1] We confine attention to binary structures here solely for the purposes of notational simplicity. By complicating the symbolism the results given here go through for structures with any number of finitary relations.

The following is proved in Dickmann (1975):

Proposition 5.6 $\mathfrak{A} \equiv_{\infty\omega} \mathfrak{B}$ *if and only if* $\mathfrak{A} \cong_{\mathbf{part}} \mathfrak{B}$.

We are going to show that, for structures, Boolean isomorphism and $\infty\omega$-equivalence coincide. To do this we first formulate a necessary and sufficient condition on a complete Boolean algebra C, similar to that of Theorem 5.1, for two structures to be C-isomorphic.

Theorem 5.7 *Let C be a complete Boolean algebra, $\mathfrak{A} = \langle A, R \rangle$ and $\mathfrak{B} = \langle B, S \rangle$ binary structures. Then the following conditions are equivalent:*

(i) *$\mathfrak{A}$ and $\mathfrak{B}$ are C-isomorphic;*

(ii) *there exists a subset $\{u_{ab} : a \in A, b \in B\}$ of C satisfying the conditions:*

    (a) $\bigvee_{b \in B} u_{ab} = 1$ *for each $a \in A$;*

    (b) $\bigvee_{a \in A} u_{ab} = 1$ *for each $b \in B$;*

    (c) $u_{ab} \wedge u_{a'b} = 0$ *whenever $a \neq a'$;*

    (d) $u_{ab} \wedge u_{ab'} = 0$ *whenever $b \neq b'$;*

    (e) *if either $(aRa'$ and $\neg bSb')$ or $(\neg aRa'$ and $bSb')$, then $u_{ab} \wedge u_{a'b'} = 0$.*

Proof (i) $\rightarrow$ (ii). Assume (i) and, using the Maximum Principle, choose $f \in V^{(C)}$ to satisfy $V^{(C)} \models f : \hat{\mathfrak{A}} \cong \hat{\mathfrak{B}}$. For $a \in A, b \in B$, define $u_{ab} \in C$ by $u_{ab} = [\![f(\hat{a}) = \hat{b}]\!]$. As in the proof of Theorem 5.1, it is easily shown that the u_{ab} satisfy (a)–(d). To prove (e), suppose that aRa' and $\neg bSb'$. Then $[\![\hat{a}\hat{R}\hat{a}']\!] = 1$ and $[\![\hat{b}\hat{S}\hat{b}']\!] = 0$, and, since $V^{(C)} \models f : \hat{\mathfrak{A}} \cong \hat{\mathfrak{B}}, [\![\hat{a}\hat{R}\hat{a}']\!] \leq [\![f(\hat{a})\hat{S}f(\hat{a}')]\!]$, so that $[\![f(\hat{a})\hat{S}f(\hat{a}')]\!] = 1$. It follows that

$$u_{ab} \wedge u_{a'b'} = u_{ab} \wedge u_{a'b'} \wedge [\![f(\hat{a})\hat{S}f(\hat{a}')]\!]$$

$$= [\![f(\hat{a}) = \hat{b} \wedge f(\hat{a}') = \hat{b}' \wedge f(\hat{a}')\hat{S}f(\hat{a}')]\!]$$

$$\leq [\![\hat{b}\hat{S}\hat{b}']\!]$$

$$= 0.$$

The argument in the case that $\neg aRa'$ and bSb' is similar.

(ii) $\rightarrow$ (i). Assume (ii), and define $f \in V^{(C)}$ by $\mathrm{dom}(f) = \{\langle \hat{a}, \hat{b} \rangle^{(C)} : a \in A, b \in B\}$ and $f(\langle \hat{a}, \hat{b} \rangle^{(C)}) = u_{ab}$. Then $u_{ab} = [\![f(\hat{a}) = \hat{b}]\!]$. Using (a)–(d) it is easy to verify that $V^{(C)} \models f$ *is a bijection of* $\hat{A}$ *and* $\hat{B}$. Now suppose that aRa'; then, using (a) and (e), and noting that $bSb' \leftrightarrow [\![\hat{b}\hat{S}\hat{b'}]\!] = 1$, we get

$$1 = \bigvee_{b \in B} u_{ab} \wedge \bigvee_{b' \in B} u_{a'b'} = \bigvee_{\substack{b,b' \in B \\ bSb'}} (u_{ab} \wedge u_{a'b'})$$

$$= \bigvee_{b,b' \in B} (u_{ab} \wedge u_{a'b'} \wedge [\![\hat{b}\hat{S}\hat{b'}]\!])$$

$$= \bigvee_{b,b' \in B} [\![f(\hat{a}) = \hat{b} \wedge f(\hat{a'}) = \hat{b'} \wedge \hat{b}\hat{S}\hat{b'}]\!]$$

$$= [\![\exists xy \in \hat{B}[f(\hat{a}) = x \wedge f(\hat{a'}) = y \wedge x\hat{S}y]]\!]$$

$$= [\![f(\hat{a})\hat{S}f(\hat{a'})]\!].$$

It follows that $[\![\hat{a}\hat{R}\hat{a'}]\!] \leq [\![f(\hat{a})\hat{S}f(\hat{a'})]\!]$, for arbitrary $a, a' \in A$, whence

$$[\![\forall xy \in A[x\hat{R}y \to f(x)\hat{S}f(y)]]\!] = \bigwedge_{a,a' \in A} [\![\hat{a}\hat{R}\hat{a'} \Rightarrow f(\hat{a})\hat{S}f(\hat{a'})]\!]$$

$$= 1.$$

Similarly, using (b) and (e), we obtain $[\![\forall xy \in A[f(x)\hat{S}f(y) \to x\hat{R}y]]\!] = 1$. It follows that $V^{(C)} \models f : \mathfrak{A} \cong \mathfrak{B}$. $\qquad\square$

As a consequence we obtain.

Corollary 5.8 *If* $\mathfrak{A} \cong_{\mathbf{part}} \mathfrak{B}$, *then* $\mathfrak{A} \cong_{\mathbf{Bool}} \mathfrak{B}$.

Proof Suppose that $P : \mathfrak{A} \cong_{\mathbf{part}} \mathfrak{B}$. Assign P the topology[2] whose basic open sets are those of the form $\{f \in P : p \subseteq f\}$ for $p \in P$. Let C be the regular open algebra of P; for each $a \in A$, $b \in B$ let $V_{ab} = \{f \in P : f(a) = b\}$ and

$$U_{ab} = \{q \in P : \forall p \supseteq q \exists f \in V_{ab}(p \subseteq f)\}.$$

Then U_{ab} is the interior of the closure of V_{ab} in P, and so $U_{ab} \in C$. From the fact that P is a partial isomorphism it is now not difficult to deduce that the U_{ab} satisfy conditions (a)–(e) of Theorem 5.7, from which it follows that $\mathfrak{A} \cong_{\mathbf{Bool}} \mathfrak{B}$. $\qquad\square$

From this and Proposition 5.6 we obtain:

Corollary 5.9 *If* $\mathfrak{A} \equiv_{\infty\omega} \mathfrak{B}$, *then* $\mathfrak{A} \cong_{\mathbf{Bool}} \mathfrak{B}$.

---

[2]This topology is the order topology (*cf.* Chapter 2) associated with the partial ordering of inverse inclusion on P.

There are now two ways of proving the converse to Corollary 5.9. We can either proceed directly, or, alternatively, prove the converse to Corollary 5.8. and then invoke Proposition 5.6. Both approaches have points of interest.

Theorem 5.10 *If $\mathfrak{A} \cong_{\mathbf{Bool}} \mathfrak{B}$, then $\mathfrak{A} \equiv_{\infty\omega} \mathfrak{B}$.*

Proof Let $\varphi(y, z)$ be a set-theoretic formula expressing the condition:

y is a binary structure, z is a sentence of $\mathscr{L}_{\infty\omega}$ and z holds in y.

The definition of satisfaction being recursive, there are restricted formulas $\psi(x, y, z)$ and $\theta(x, y, z)$ such that φ is equivalent in ZF both to $\exists x\psi$ and $\forall x\theta$.

Now suppose that $\mathfrak{A} \cong_{\mathbf{Bool}} \mathfrak{B}$; let C be a complete Boolean algebra such that $\mathfrak{A}$ and $\mathfrak{B}$ are C-equivalent. Let σ be a sentence of $\mathscr{L}_{\infty\omega}$ for which $\mathfrak{A} \models \sigma$. Then $\varphi(\mathfrak{A}, \sigma)$ holds in V, and hence there is $a \in V$ for which $\psi(a, \mathfrak{A}, \sigma)$ holds in V. Since ψ is restricted, it follows from Theorem 1.23(v) that $V^{(C)} \models \psi(\hat{a}, \hat{\mathfrak{A}}, \hat{\sigma})$. Therefore $V^{(C)} \models \exists x\psi(x, \hat{\mathfrak{A}}, \hat{\sigma})$, whence $V^{(C)} \models \varphi(\hat{\mathfrak{A}}, \hat{\sigma})$. Therefore $V^{(C)} \models \varphi(\hat{\mathfrak{B}}, \hat{\sigma})$, since $\hat{\mathfrak{A}}$ and $\hat{\mathfrak{B}}$ are isomorphic in $V^{(C)}$ and so satisfy the same sentences. It follows that $V^{(C)} \models \forall x\theta(x, \hat{\mathfrak{B}}, \hat{\sigma})$, and so in particular $V^{(C)} \models \theta(\hat{a}, \hat{\mathfrak{B}}, \hat{\sigma})$ for every $a \in V$. Since θ is restricted, we infer that $\theta(a, \mathfrak{B}, \sigma)$ holds in V for every $a \in V$, that is, $\forall x\theta(x, \mathfrak{B}, \sigma)$ holds in V. Accordingly $\varphi(\mathfrak{B}, \sigma)$ holds in V, i.e. $\mathfrak{B} \models \sigma$. Replacing σ by $\neg\sigma$ in this argument yields the reverse implication; we conclude that $\mathfrak{A} \equiv_{\infty\omega} \mathfrak{B}$, and the theorem is proved. $\qquad\square$

Theorem 5.11 *If $\mathfrak{A} \cong_{\mathbf{Bool}} \mathfrak{B}$, then $\mathfrak{A} \cong_{\mathbf{part}} \mathfrak{B}$.*

Proof Suppose that $\mathfrak{A} \cong_{\mathbf{Bool}} \mathfrak{B}$; then by Theorem 5.7 there is a complete Boolean algebra C satisfying conditions (a)–(e) of that theorem. For each $0 \neq u \in C$ put

$$\tilde{v} = \{\langle a, b\rangle \in A \times B : v \leq u_{ab}\}.$$

We claim that

$$P = \{\tilde{v} : 0 \neq v \in C\}$$

is a partial isomorphism between $\mathfrak{A}$ and $\mathfrak{B}$.

(1) *Each $\tilde{v} \in P$ is a one–one function.* For if $\langle a, b\rangle \in \tilde{v}$ and $\langle a, b'\rangle \in \tilde{v}$, then $v \leq u_{ab} \wedge u_{ab'} = 0$ if $b \neq b'$ by Theorem 5.7(d); so since $v \neq 0$ we must have $b = b'$. Accordingly $\tilde{v}$ is a function. In a similar way, now using Theorem 5.7(c), we can show that $\tilde{v}$ is one–one.

(2) *Each $\tilde{v} \in P$ is an isomorphism of its domain onto its range.* For suppose $\langle a, b\rangle \in \tilde{v}, \langle a', b'\rangle \in \tilde{v}$ and aRa'. Then if $\neg bSb'$, it follows from Theorem 5.7(e) that $v \leq u_{ab} \wedge u_{a'b'} = 0$, so since $v \neq 0$ we must have bSb'. Similarly, if bSb', we obtain aRa'.

(3) Suppose $\tilde{v} \in P$ and $a \in A$. By Theorem 5.7(a), we have $\bigvee_{b \in B} u_{ab} = 1$; hence

$$v = v \wedge \bigvee_{b \in B} u_{ab} = \bigvee_{b \in B} v \wedge u_{ab}.$$

Since $v \neq 0$, for some $b \in B$ we must have $w = v \wedge u_{ab} \neq 0$. Then $\tilde{w} \in P, \tilde{v} \subseteq \tilde{w}$ and $\langle a, b \rangle \in \tilde{w}$, whence $a \in \mathrm{dom}(\tilde{w})$.

Similarly, now using Theorem 5.7(b), we get $\tilde{w} \in P$ such that $\tilde{v} \subseteq \tilde{w}$ and $b \in \mathrm{ran}(\tilde{w})$.

Accordingly P is a partial isomorphism between $\mathfrak{A}$ and $\mathfrak{B}$ and the proof is complete. $\qquad\square$

All this gives the promised

Theorem 5.12 $\mathfrak{A} \cong_{\mathbf{Bool}} \mathfrak{B}$ *if and only if* $\mathfrak{A} \equiv_{\infty\omega} \mathfrak{B}$.

Finally, let us call a class of sentences of $\mathscr{L}_{\infty\omega}$ *Boolean categorical* if it has a model and any pair of its models are Boolean isomorphic. It follows immediately from Theorem 5.12 that *the* $\infty\omega$-*theory of any structure is Boolean categorical.*

Applications to the theory of Boolean algebras

It is a fact that any countably generated Boolean algebra is a homomorphic image of the free Boolean algebra on countably many generators. (This algebra may be explicitly described as the algebra of clopen subsets of the Cantor ternary set with its usual topology.) In 1964 Gaifman and Hales (independently) showed that the situation in respect of *complete* Boolean algebras is strikingly different.

Let us say that a complete Boolean algebra is *countably completely generated* (ccg) if there is a countable subset X of B such that the least complete subalgebra of B, which includes X is B itself. (Under these conditions X is said to *completely generate* B.) Now Gaifman and Hales showed that there are ccg complete Boolean algebras of *arbitrarily high cardinality*: this implies at once that there is no Boolean algebra B such that each ccg complete Boolean algebra is a homomorphic image of B.

In 1965 Solovay used the properties of collapsing algebras to provide a remarkably simple proof of Gaifman and Hales' theorem. Essentially, Solovay observed that if $B = \mathrm{RO}(\lambda^{\omega})$, then $V^{(B)}$ can be obtained by adjoining a B-valued set E of natural numbers to V, and that the Boolean values $[\![\hat{\eta} \in E]\!]$ completely generate B: *cf.* Problems 4.35 and 4.36.

Theorem 5.13 *Let* λ *be an infinite cardinal and put* $B = \mathrm{RO}(\lambda^{\omega})$. *Then* B *is* ccg *and* $|B| \geq \lambda$. *Hence there are* ccg *complete Boolean algebras of arbitrarily high cardinality.*

Proof Define $f \in V^{(B)}$ by

$$\mathrm{dom}(f) = \{\langle \hat{m}, \hat{\eta} \rangle^{(B)} : \langle m, n \rangle \in \omega \times \lambda\}$$

and

$$f(\hat{m}, \hat{\eta} \rangle^{(B)}) = \{g \in \lambda^{\omega} : g(m) = \eta\}.$$

One can then show, as in the proof of Theorem 5.1, that f is collapsing function from $\hat{\omega}$ to $\hat{\lambda}$ in $V^{(B)}$, that is,

$$V^{(B)} \models f \ is \ a \ map \ of \ \hat{\omega} \ onto \ \hat{\lambda}. \tag{1}$$

Let $b_{m\eta} = f(\langle \hat{m}, \hat{\eta} \rangle^{(B)})$; it is then not hard to verify that

$$b_{m\eta} = [\![f(\hat{m}) = \hat{\eta}]\!].$$

The $b_{m\eta}$ form a subbase for the product topology on λ^{ω}, and using this fact it is straightforward to show that the $b_{m\eta}$ completely generate B. Since there are λ different $b_{m\eta}$, it follows that $|B| \geq \lambda$.

Put $a_{mn} = [\![f(\hat{m}) < f(\hat{n})]\!]$, for $m, n \in \omega$. We shall show that the a_{mn} completely generate B, thereby proving the theorem. Since the b_{mn} completely generate B, it suffices to show that each $b_{m\eta}$ is in the complete subalgebra B' of B completely generated by the a_{mn}.

We prove this last assertion by induction on η. Suppose that, for all $\xi < \eta$ and all $m \in \omega$ we have $b_{m\xi} \in B'$. We show that both $[\![f(\hat{m}) < \hat{\eta}]\!]$ and $[\![f(\hat{m}) \leq \hat{\eta}]\!]$ are in B', for all $m \in \omega$. We shall then have

$$b_{m\eta} = [\![f(\hat{m}) = \hat{\eta}]\!] = [\![f(\hat{m}) \leq \hat{\eta}]\!] \wedge [\![f(\hat{m}) < \hat{\eta}]\!]^* \in B',$$

completing the induction step.

First we have

$$[\![f(\hat{m}) < \hat{\eta}]\!] = \bigvee_{\xi < \eta} [\![f(\hat{m}) = \hat{\xi}]\!] = \bigvee_{\xi < \eta} b_{m\xi} \in B', \tag{2}$$

since, by inductive hypothesis, $b_{m\xi} \in B'$ for all $\xi < \eta$. And finally,

$$[\![f(\hat{m}) \leq \hat{\eta} = [\![\forall \alpha < f(\hat{m})(\alpha < \hat{\eta})]\!]$$
$$= [\![\forall x \in \hat{\omega}[f(x) < f(\hat{m}) \to f(x) < \hat{\eta}]]\!] \quad (\text{using}(1))$$

$$= \bigwedge_{n \in \omega} [\![f(\hat{n}) < f(\hat{m})]\!] \Rightarrow [\![f(\hat{n}) < \hat{\eta}]\!]]$$

$$= \bigwedge_{n < \omega} [a_{nm} \Rightarrow [\![f(\hat{n}) < \hat{\eta}]\!] \in B' \quad \text{(by (2))}. \qquad \square$$

In 1967 Kripke strengthened Solovay's result by showing that collapsing algebras enjoy a remarkable *embedding* property. If there exists a complete monomorphism of a Boolean algebra A into B, let us say that A can be *completely embedded* in B. Then Kripke's result is:

Theorem 5.14 *Let A be a Boolean algebra of infinite cardinality κ. Then A can be completely embedded in the collapsing $(\aleph_0, 2^\kappa)$-algebra.*

Proof Let $\lambda = 2^\kappa$ and let $B = \mathrm{RO}(\lambda^\omega)$ be the collapsing $(\aleph_0, \lambda)$-algebra. We know from Corollary 5.2 that $V^{(B)} \models \hat{\lambda}$ *is countable*, whence

$$V^{(B)} \models (PA)\hat{\ } \text{ is countable.}$$

Let $\{Q_\xi : \xi < \kappa\}$ be a partition of λ and for each $\xi < \kappa$ put

$$b_\xi = \{f \in \lambda^\omega : f(0) \in Q_\xi\}.$$

Then the b_ξ form a partition of unity in B. Let $\{a_\xi : \xi < \kappa\}$ be an enumeration of $A - \{0_A\}$. Then by the Mixing Lemma, there is $b \in V^{(B)}$ such that $b_\xi \leq [\![b = \hat{a}_\xi]\!]$ for all $\xi < \kappa$. It follows at once that

$$[\![b \in \hat{A}]\!] \geq \bigvee_{\xi < \kappa} [\![b = \hat{a}_\xi]\!] \geq \bigvee_{\xi < \kappa} b_\xi = 1.$$

The predicate 'x is a Boolean algebra' is a restricted formula, so that

$$V^{(B)} \models \hat{A} \text{ is a Boolean algebra.}$$

Moreover we have, for each $\xi < \kappa$,

$$b_\xi \leq [\![b = \hat{a}_\xi]\!] = [\![b = \hat{a}_\xi]\!] \wedge [\![\hat{a}_\xi \neq \hat{0}_A]\!] \leq [\![b \neq \hat{0}_A]\!] = [\![b \neq 0_{\hat{A}}]\!],$$

so that

$$1 = \bigvee_{\xi < \kappa} b_\xi \leq [\![b \neq 0_{\hat{A}}]\!].$$

Let $S = \{X \in PA : \bigvee X$ exists in $A\}$. Then $V^{(B)} \models \hat{S}$ *is countable*, and since $V^{(B)} \models$ *Rasiowa–Sikorski Lemma*,

$$V^{(B)} \models \exists U[U \text{ is an } \hat{S}\text{-complete ultrafilter in } \hat{A} \text{ and } b \in U].$$

The Maximum Principle now implies the existence of a $U \in V^{(B)}$ such that

$$V^{(B)} \models U \text{ is an } \hat{S}\text{-complete ultrafilter in } A \text{ and } b \in U. \tag{1}$$

We define $h : A \to B$ by

$$h(a) = [\![\hat{a} \in U]\!]$$

for $a \in A$. It is easy to verify that h is a homomorphism of A into B. To see that h is complete, observe that, if $X \in S$ and $a = \bigvee X$ in A, then $[\![\hat{a} = \bigvee \hat{X} \text{ in } \hat{A}]\!] = 1$, so that, using (1),

$$h(\bigvee X) = h(a) = [\![\hat{a} \in U]\!] = [\![\bigvee \hat{X} \in U]\!]$$

$$= [\![\exists x \in \hat{X}(x \in U)]\!] = \bigvee_{x \in X} [\![\hat{x} \in U]\!]$$

$$= \bigvee_{x \in X} h(x) = \bigvee h[X].$$

And finally h is one–one, because if $0_A \neq a \in A$, then $a = a_\xi$ for some $\xi < \kappa$, whence $h(a) = [\![\hat{a}_\xi \in U]\!] \geq [\![b \in U]\!] \wedge [\![\hat{a}_\xi = b]\!] = [\![\hat{a}_\xi = b]\!] \neq 0$. $\square$

Theorems 5.13 and 5.14 immediately give

Corollary 5.15 *Each Boolean algebra can be completely embedded in a* ccg *complete Boolean algebra.*

Problems

5.16 (Universal complete Boolean algebras) Let κ be an infinite cardinal. A Boolean algebra B is said to be κ-*universal* if for each Boolean algebra A of cardinality $\leq \kappa$ there is a monomorphism of A into B. If B is *complete*, show that the following conditions are equivalent:

(i) B is κ-universal;

(ii) B has an antichain of cardinality κ. (For (ii) $\to$ (i); argue as in the proof of Theorem 5.14, ignoring the collapsing property.)

5.17 (Homogeneous Boolean algebras) Show that, for each λ, the collapsing $(\aleph_0, \lambda)$-algebra is homogeneous. Deduce that each Boolean algebra can be completely embedded in a homogeneous complete Boolean algebra. (To establish homogeneity, argue along the lines of Lemma 3.7.)

6

ITERATED BOOLEAN EXTENSIONS, MARTIN'S AXIOM, AND SOUSLIN'S HYPOTHESIS

Souslin's hypothesis

It is a fact that the real line can be characterized up to order isomorphism as the unique linearly ordered set, which is order dense, complete and unbounded, as well as *separable*, that is, has a countable subset intersecting each nonempty open interval. In 1920 Souslin raised the question as to whether separability could be replaced by the following—apparently weaker—condition:

$$\textit{Every family of disjoint open intervals is countable.} \tag{$*$}$$

Souslin's problem may be equivalently stated in the form of

Souslin's hypothesis (SH) *Every order dense linearly ordered set satisfying* $(*)$ *above is separable.*

To see that the two formulations are equivalent, let us write SH$'$ for the assertion that any complete, unbounded, order dense, linearly ordered set satisfying $(*)$ is isomorphic to the real line R. Then clearly SH $\rightarrow$ SH$'$. Conversely, if SH$'$ holds, let P be an (infinite) order dense linearly ordered set satisfying $(*)$. Then P contains a copy of the ordered set Q of rational numbers. The (Dedekind) order completion C of P (with its end-points removed) is then order dense, unbounded, and satisfies $(*)$ since P does. Thus SH$'$ implies that C is isomorphic to R. So P may be regarded as a dense subset of R, which contains Q. Clearly P is then separable, and SH follows.

In this chapter we are going to establish first the independence and then the relative consistency of SH with ZFC. The first step in this process is to introduce the notion of a tree.

A *tree* is a partially ordered set $\langle T, \leq_T \rangle$ with the property that, for each $x \in T$, the set $\{y : y <_T x\}$ of predecessors of x is well-ordered by $\leq_T$. For each $x \in T$, we write $o(x)$ for the order type of $\{y : y <_T x\}$; $o(x)$ is then an ordinal. For each ordinal α, the αth *level* of T consists of all $x \in T$ for which $o(x) = \alpha$. The *height* of T is the least α such that the αth level of T is empty. A *branch* in T is a maximal linearly ordered subset of T. A subset X of T is said to be *free* if any two different elements x, y of X are incomparable, that is, neither $x \leq_T y$

nor $y \leq_T x$. Finally, a tree T is called a *Souslin tree* if

> T has height ω_1;
> every branch in T is countable;
> every free subset of T is countable.

We can now reformulate Souslin's hypothesis in terms of Souslin trees.

Lemma 6.1 SH *holds iff there are no Souslin trees.*

Proof Sufficiency Suppose SH fails; then there is a dense linearly ordered set P which is not separable but in which every family of disjoint open intervals is countable. We use P to construct a Souslin tree T as follows. T will consist of closed (nondegenerate) intervals in P, and will be partially ordered by *inverse inclusion* $\supseteq$.

We construct T by recursion on $\alpha < \omega_1$. Let $I_0 = [a_0, b_0]$ be arbitrary (with $a_0 < b_0$). Assuming that we have got all I_β for $\beta < \alpha$, consider the countable set $C = \{a_\beta : \beta < \alpha\} \cup \{b_\beta : \beta < \alpha\}$ of endpoints of the intervals I_β. Since P is not separable, there must be an interval disjoint from C; let $I_\alpha = [a_\alpha, b_\alpha]$ be one. The set $T = \{I_\alpha : \alpha < \omega_1\}$ is uncountable and partially ordered by $\supseteq$. If $\alpha < \beta$, then either $I_\alpha \supseteq I_\beta$ or $I_\alpha \cap I_\beta = \emptyset$. It follows that, for each α, the set $\{I \in T : I \supseteq I_\alpha\}$ is well-ordered by $\supseteq$ and so T is a tree.

We show that T has no uncountable branches and no uncountable free subsets. Clearly the height of T then cannot exceed ω_1; and since every level of T is evidently free and T is uncountable, it follows that T has height ω_1.

If I, J are incomparable members of T, then they are, by construction, disjoint intervals of P; so any free subset of T is countable. To show T has no uncountable branches, we observe that if b is a branch of length ω_1, then the left endpoints of the intervals $I \in b$ of from an increasing sequence $\{x_\alpha : \alpha < \omega_1\}$ of points of P. But then the intervals $(x_\alpha, x_{\alpha+1}), \alpha < \omega_1$ form an uncountable collection of disjoint open intervals in P, contradicting assumption.

Necessity Let T be a Souslin tree. First we remove from T all points $x \in T$ such that $\{y \in T : x \leq y\}$ is countable, thus obtaining a new tree T'. It is easy to see that for each $x \in T'$ there exists $y \in T', y > x$, at each greater level $< \omega_1$. Next, we discard from T' all points $x \in T'$ for which there is only *one* point $y > x$ at the next level. This gives us a new tree T''. Finally from T'' we expunge all points except those at *limit* levels. This yields a Souslin tree in which each level has cardinality $\aleph_0$. Thus, without loss of generality we may assume that the same holds for T.

Now let P be the set of all branches in T; we order P as follows. First, we order each level of T as in the rational numbers. Then, given $b_1, b_2 \in P$, we put $b_1 < b_2$ if the αth element of b_1 precedes the αth element of b_2 in the ordering of U_α, where U_α is the least level at which b_1 and b_2 differ. Clearly this prescription makes P linearly ordered and dense.

If (a, b) is an interval in P, it is not hard to see that there is $x \in T$ such that $I_x \subseteq (a, b)$, where I_x is the interval $I_x = \{c \in P : x \in c\}$. Moreover, if $I_x \cap I_y = \emptyset$, then x and y are incomparable points of T. It follows that every collection of disjoint open intervals in P is countable.

On the other hand, P is not separable. For if C is a countable set of branches in T, let α be a countable ordinal exceeding the length of any branch in C. Then if x is any point at level α, the interval I_x is disjoint from C. $\qquad\square$

The independence of SH

Before we begin the proof of independence of SH from ZFC, we require a combinatorial lemma.

Lemma 6.2 *Let S be an uncountable collection of finite sets. Then there is an uncountable $Z \subseteq S$ and a finite set A such that $X \cap Y = A$ for any distinct elements $X, Y \in Z$.*

Proof Since S is uncountable, it is clear that uncountably many $X \in S$ have the same cardinality. Thus we may assume that, for some $n, |X| = n$ for all $X \in S$. The lemma is now proved by induction on n. If $n = 1$, the lemma is trivial. So assume its truth for n, and let S be such that $|X| = n + 1$ for all $X \in S$.

Case 1 Some element a belongs to uncountably many $X \in S$. In this case we obtain the required $Z \subseteq S$ by applying the inductive hypothesis to the family $\{X - \{a\} : X \in S \wedge a \in X\}$.

Case 2 Not case 1. Here we can easily construct a disjoint family $Z = \{X_\alpha : \alpha < \omega_1\} \subseteq S$ by choosing inductively X_α to be a member of S disjoint from all X_ξ for $\xi < \alpha$. $\qquad\square$

Now let M be any countable transitive $\in$-model of ZFC. We shall show that M has a generic extension $M[G]$ containing a Souslin tree.

In M, let P be the set of all finite trees $\langle T, \leq_T \rangle$ such that $T \subseteq \omega_1$ and $\alpha \leq_T \beta \to \alpha \leq \beta$ (as ordinals). We partially order P by stipulating that:

$$\langle T_1, \leq_{T_1} \rangle \preceq \langle T_2, \leq_{T_2} \rangle \leftrightarrow T_1 \supseteq T_2 \wedge \leq_{T_2} \, = \, \leq_{T_1} | T_2.$$

In the sequel we shall usually write '$\leq_1$' for '$\leq_{T_1}$', etc.

A partially ordered set is said to satisfy the *countable chain condition* (ccc) if every subset consisting of incompatible elements is countable. It is clear that a partially ordered set satisfies ccc in this sense iff its Boolean completion (*cf.* Problem 2.4) satisfies the ccc in the sense of Chapter 1.

Lemma 6.3 $\langle P, \preceq \rangle$ *satisfies* ccc.

Proof Let S be an uncountable subset of P. Using Lemma 6.2, we obtain an uncountable subset $Z_1 \subseteq S$ and a finite set $A \subseteq \omega_1$ such that, for any distinct

$T_1, T_2 \in Z_1$ we have $T_1 \cap T_2 = A$ and $\leq_1 |A = \leq_2| A$. Now discard from Z_1 all trees T for which there exist $\alpha \in A$ and $\beta < \alpha$ such that $\tilde{\beta} \in T - A$. Only countably many trees are lost from Z_1 in this way. If we call what is left Z_2, then Z_2 is uncountable. But now any $T_1, T_2 \in Z$ are compatible: for if we define $\leq_3$ on $T_3 = T_1 \cup T_2$ by

$$\alpha \leq_3 \beta \leftrightarrow \alpha \leq_1 \beta \vee \alpha \leq_2 \beta,$$

then $\langle T_3, \leq_3 \rangle \preceq \langle T_1, \leq_1 \rangle$ and $\langle T_3, \leq_3 \rangle \preceq \langle T_2, \leq_2 \rangle$. The lemma follows. $\qquad \square$

Now, using Problem 4.34, we choose an M-generic subset G of P and put $H = \bigcup G$. We let $\leq_H$ be the partial ordering on H induced in the obvious way by the partial orderings of the members of G. It is easy to see that H is a tree in $M[G]$.

Theorem 6.4 $M[G] \models H$ *is a Souslin tree.*

Proof Let us call a point x of a tree a *branch point* if there are at least two points $y > x$ at the next level above x.

We claim first that:

(1) every point of H is a branch point;

(2) for each $\alpha \in H$, the set $\{\beta \in H : \alpha <_H \beta\}$ is uncountable in $M[G]$.

To prove (1), we take any $\alpha \in H$, choose $T_0 \in G$ such that $\alpha \in T_0$ and observe that the set of $T \in P$ in which α is a branch point is dense below T_0 in P. Consequently there must be $T \in G$ is which α is a branch point; it follows that α is a branch point in H.

To prove (2), we again take $\alpha \in H$ and $T_0 \in G$ such that $\alpha \in T_0$. Next we observe that, for each $\xi < \omega_1^{(M)}$, the set

$$Q = \{T \in P : \alpha \in T \wedge \exists \beta \in T[\xi < \beta \wedge \alpha <_T \beta]\}$$

is dense below T_0. Since G is generic, $G \cap Q \neq \emptyset$. Hence for each $\xi < \omega_1^{(M)}$ there is $\beta \in H$ such that $\xi < \beta$ and $\alpha <_H \beta$. But Lemma 6.3 implies that $\omega_1^{(M)} = \omega_1^{(M[G])}$, and (2) follows.

Now we can prove the following.

(3) $M[G] \models$ *every free subset of H is countable.*

For suppose (3) is false; then, for some $E \in M[G]$ we have

E is an uncountable free subset of H.

Let B be the Boolean completion of P, let i be the canonical map of $M^{(B)}$ onto $M[G]$, and let $\tilde{E}, \tilde{H} \in M^{(B)}$ be such that $i(\tilde{E}) = E, i(\tilde{H}) = H$. Then, for some $T_0 \in G$ we have

$$T_0 \Vdash \tilde{E} \text{ is an uncountable free subset of } \tilde{H}. \qquad (*)$$

It follows that for each $\xi < \omega_1^{(M)}$ we can find $T \in P$ and $\alpha_T \in T$ such that $T \preceq T_0$, $\xi < \alpha_T$ and $T \Vdash \hat{\alpha}_T \in \tilde{E}$. In this way we obtain an uncountable (in M) set $S \subseteq P$ and, for each $T \in S$ an ordinal $\alpha_T \in T$ such that $T \Vdash \hat{\alpha}_T \in \tilde{E}$. Lemma 6.2 now yields an uncountable subset Z_1 of S and a finite set A such that, for any $T_1 \neq T_2$ in Z_1 we have $T_1 \cap T_2 = A$ and $\leq_{T_1} |A =<_{T_2} |A$. Without loss of generality we may assume that $\alpha_T \notin A$ for any $T \in Z_1$. Since A is finite, there must be an uncountable subset Z_2 of Z such that, for any $T_1, T_2 \in Z_2$,

$$A \cap \{\beta : \beta <_{T_1} \alpha_{T_1}\} = A \cap \{\beta : \beta <_{T_2} \alpha_{T_2}\}.$$

From Z_2 we discard all trees T for which there is $\alpha \in A$ and $\beta < \alpha$ such that $\beta \in T - A$. Only countably many trees are lost in this way, so if we call what is left Z_3, then Z_3 is uncountable.

If $T_1, T_2 \in Z_3$ then we have $\langle T_3, \leq_3 \rangle \preceq \langle T_1, \leq_1 \rangle$ and $\langle T_3, \leq_3 \rangle \preceq \langle T_2, \leq_2 \rangle$ where $T_3 = T_1 \cup T_2$ and $\leq_3$ is defined as follows: if $\alpha_{T_1} < \alpha_{T_2}$, then

$$\alpha \leq_3 \beta \leftrightarrow \alpha \leq_1 \beta \vee \alpha \leq_2 \beta$$
$$\vee \, [\alpha \leq_1 \alpha_{T_1} \wedge \exists \gamma \leq_2 \alpha_{T_2} (\alpha \leq \gamma \wedge \gamma \leq_2 \beta)]$$
$$\vee \, [\alpha \leq_2 \alpha_{T_2} \wedge \exists \gamma \leq_1 \alpha_{T_1} (\alpha \leq_\gamma \gamma \wedge \gamma \leq_1 \beta)]$$
$$\vee \, [\alpha \leq_1 \alpha_{T_1} \wedge \alpha_{T_2} \leq_2 \beta],$$

and similarly when $\alpha_{T_2} < \alpha_{T_1}$.

It is clear that α_{T_1} and α_{T_2} are comparable with respect to $\leq_{T_3}$. So

$$T_3 \Vdash \hat{\alpha}_{T_1} \in \tilde{E} \wedge \hat{\alpha}_{T_2} \in \tilde{E} \wedge \hat{\alpha}_{T_1} \text{ is comparable in } \tilde{H} \text{ with } \hat{\alpha}_{T_2}.$$

But this contradicts $(*)$. This proves (3).

It therefore remains to prove the following.

(4) $M[G] \models H$ *has height* ω_1;

(5) $M[G] \models$ *every branch in H is countable.*

To prove (4), we suppose that (in $M[G]$) H has height $< \omega_1$. Then for any $\alpha \in H$ the set $\{o(\beta) : \alpha <_H \beta\}$ is countable and it follows from (2) that for some $\gamma < \omega_1$ the set $\{\beta \in H : o(\beta) = \alpha\}$ is uncountable. But this latter set is free, contradicting (3).

Finally, suppose (5) is false; let b be a branch in H of length ω_1 (in $M[G]$). Using (1) we choose for each $x \in b$ an element $f(x) \geq_H x$ not in b. Then $\{f(x) : x \in b\}$ is free and uncountable in $M[G]$, contradicting (3). $\qquad \square$

Corollary 6.5 *If* ZF *is consistent, so are* ZFC + GCH + ¬SH *and* ZFC+ ¬CH + ¬SH.

Proof Let B be the Boolean completion of P, and assume that $M \models$ GCH. Since P satisfies ccc, so does B and it is easily shown, using GCH in M, that $M \models |B| = 2^{\aleph_0}$. It follows from Theorems 2.8 and 6.4 that $M[G] \models$ GCH $\wedge \neg$SH. Since this holds for an arbitrary generic set G is P, we infer that $M[U] \models$ GCH $\wedge \neg$SH for an arbitrary generic ultrafilter U in B and hence, using Problem 4.26, that $M^{(B)} \models$ GCH $\wedge \neg$SH. By the remarks following Lemma 4.25, we can transfer this to $V^{(B)}$, obtaining (now under that assumption that GCH holds in V) that $V^{(B)} \models$ GCH $\wedge \neg$SH. The first assertion now follows from Theorem 1.19.

If on the other hand $M \models 2^{\aleph_0} \geq \aleph_2$, then $M[G] \models \aleph_2^{(M)} \leq 2^{\aleph_0}$. But B satisfies ccc, and therefore, by Theorem 1.51, we have $M[G] \models \aleph_2^{(M)} = \aleph_2$. Hence $M[G] \models \aleph_2 \leq 2^{\aleph_0}$, yielding the second assertion as above. $\square$

Martin's axiom

Having established the independence of SH from ZFC, we want now to demonstrate its relative consistency. Rather than attempting to go about doing this directly, however we instead formulate a principle that easily implies the nonexistence of Souslin trees, and give a relative consistency proof for this principle. The principle, known as *Martin's axiom*, provides an interesting alternative to the continuum hypothesis, and has become an important tool in general topology and infinite combinatorics.

Let κ be an infinite cardinal. *Martin's axiom at level κ* is the assertion

MA_κ: *If B is a Boolean algebra satisfying ccc and S is any family of subsets of B (each member of which has a join) such that $|S| < \kappa$, then there is an S-complete ultrafilter in B.*

Notice that $\mathrm{MA}_{\aleph_1}$ is just a special case of the Rasiowa–Sikorski lemma, and hence provable in ZFC. So MA_κ is only novel when $\kappa > \aleph_1$. We observe, however the following.

Lemma 6.6

$$\mathrm{ZFC} \vdash \mathrm{MA}_\kappa \to \kappa \leq 2^{\aleph_0}.$$

Proof Let $B = \mathrm{RO}(2^\omega)$; then B satisfies ccc. For each $s \subseteq \omega$ let χ_s be the characteristic function of s and let $X_s = \{N(p) : p \not\subseteq \chi_s\}$. It is easily verified that $\bigvee X_s = 1$ in B; so if MA_κ with $\kappa > 2^{\aleph_0}$ there would be a $\{X_s : s \subseteq \omega\}$-complete ultrafilter U in B. Now define $t \subseteq \omega$ by

$$n \in t \leftrightarrow N(\{\langle n, 1 \rangle\}) \in U.$$

It is now easy to show that for any subset $s \subseteq \omega$ we have $t \neq s$, which is a contradiction. $\square$

Problem 6.7 (A stronger form of Martin's axiom?) Let $MA_\kappa!$ be obtained from MA_κ by dropping the restriction to algebras satisfying ccc. Show that $MA_\kappa! \to \kappa \leq \aleph_1$. (Consider the collapsing $(\aleph_0, \aleph_1)$-algebra.)

We next give several alternative formulations of MA_κ. If P is a partially ordered set, and S a family of subsets of P, a subset G of P is said to be *S-generic* (*cf.* Problem 4.34) if

(a) $x \in G, x \leq y \to y \in G$;

(b) $x, y \in G \to \exists z \in G[z \leq x \wedge z \leq y]$;

(c) $X \in S \wedge X$ dense in $P \to X \bigcap G \neq \emptyset$.

Theorem 6.8 *The following are equivalent for any infinite cardinal κ:*

(i) MA_κ.

(ii) *If B is a Boolean algebra satisfying ccc with $|B| < \kappa$ and S is a family of subsets with $|S| < \kappa$, each member of which has a join, then there is an S-complete ultrafilter in B.*

(iii) *If P is a partially ordered set satisfying ccc such that $|P| < \kappa$ and S is a family of subsets of P with $|S| < \kappa$, then there is an S-generic subset of P.*

(iv) *Same as (iii) but with '$|S| < \kappa$' omitted.*

Proof (i) $\to$ (ii) is trivial.

(ii)$\to$ (iii). Assume (ii) and let P be a partially ordered set satisfying ccc with $|P| < \kappa$. Let C be the Boolean completion of P and let $j : P \to C$ be the canonical map. Also let B be the subalgebra of C generated by $j[P]$. Then $|B| < \kappa$. Now observe that if E is any family of $< \kappa$ dense subsets of B, then $\bigvee X = 1$ in B for each $X \in E$, and therefore, by (ii), there is an E-complete ultrafilter in B. If S is any family of dense sets in P with $|S| < \kappa$, consider the following dense subsets of B : (a) all $j[X]$ with $X \in S$; (b) all $j[Z_{xy}]$ for $x, y \in P$, where

$$Z_{xy} = \{z \in P : [z \leq x \wedge z \leq y] \vee \neg\mathrm{Comp}(z, x) \vee \neg\mathrm{Comp}(z, y)\}.$$

Since $|P| < \kappa$, there are $< \kappa$ sets of the form (a) or (b); so (ii) yields an ultrafilter U in B which meets all of them. It is now readily verified that $j^{-1}[U]$ is an S-generic subset of P.

(iii) $\to$ (iv). Assume (iii) and let P be a partially ordered set satisfying ccc. Let S be a family of $< \kappa$ dense subsets of P. For each $X \in S$, let Q_X be a maximal subset of X consisting of mutually incompatible elements. Since P satisfies ccc, each Q_X is countable. Hence there is a subset Q of P of cardinality $< \kappa$ such that $Q_X \subseteq Q$ for all $X \in S$ and

$$x, y \in Q \wedge \mathrm{Comp}(x, y) \to \exists z \in Q[z \leq x \wedge z \leq y].$$

Each Q_X is a maximal incompatible subset of Q and, for each $X \in S$, $K_X = \{x \in Q : x \leq y$ for some $y \in Q_X\}$ is dense in Q.

The partially ordered set Q is of cardinality $< \kappa$ and satisfies ccc. So by (iii) there is a $\{K_X : X \in S\}$-generic subset G of Q. It is now easily verified that the set $\{x \in P : \exists y \in G[y \leq x]\}$ is S-generic in P.

(iv) $\rightarrow$ (i). Assume (iv), let B be a Boolean algebra with ccc, let S be a family of subsets of B, each member of which has a join, let $P = B - \{0\}$ and for each $X \in S$ let

$$D_X = \{y \in P : \exists x \in X[y \leq x] \vee \forall x \in X[y \wedge x = 0]\}.$$

Each D_X is dense in P and so (iv) implies that there is a $\{D_X : X \in S\}$-generic subset G of P. Clearly any ultrafilter in B extending G is S-complete. $\qquad \square$

The principle $\mathrm{MA}_{2^{\aleph_0}}$ is called *Martin's axiom* and is written simply MA. Clearly MA is a consequence of CH; but we shall see that MA can hold even when CH fails. Moreover, we can use Theorem 6.8 to show that, in this eventuality, Souslin's hypothesis holds:

Theorem 6.9

$$\mathrm{MA} + 2^{\aleph_0} > \aleph_1 \rightarrow \mathrm{SH}.$$

Proof Suppose SH is false; then there is a Souslin tree T. Let T' be the set of $x \in T$ for which $\{y \in T : x \leq_T y\}$ is uncountable. It is then easy to see that T' is a Souslin tree with the property:

for each $x \in T'$ there is some $y > x$ at each greater level $(< \omega_1)$. $\qquad (*)$

Now let P be the partially ordered set obtained from T' by reversing the order. It is easy to see that P satisfies ccc. For each $\alpha < \omega_1$ let

$$X_\alpha = \{x \in T' : o(x) > \alpha\}.$$

Using $(*)$, one verifies that X_α is dense in P.

Thus, assuming $\mathrm{MA} + 2^{\aleph_0} > \aleph_1$, there is an $\{X_\alpha : \alpha < \omega_1\}$-generic subset G of P. It is now a routine matter to verify that G is a branch in T' of cardinality ω_1, a contradiction. $\qquad \square$

It follows from Theorem 6.9 that, in order to establish the relative consistency of SH, it suffices to establish that of $\mathrm{MA} + 2^{\aleph_0} > \aleph_1$. We shall achieve this by constructing an increasing sequence of Boolean extensions of the universe in such a way that at each succesive stage a potential counterexample to Martin's axiom is 'liquidated' by adjoining an ultrafilter of the appropriate sort. Then, with some finesse, we can show that the 'limit' of this sequence of Boolean extensions (in a sense to be made precise later on) is a Boolean-valued model of $\mathrm{MA} + 2^{\aleph_0} > \aleph_1$. We turn now to elaborating this procedure, which is called the method of *iterated Boolean extensions*.

Iterated Boolean extensions

Let B be a complete Boolean algebra, and suppose we are given elements $C, \leq_C$ of $V^{(B)}$ such that

$$V^{(B)} \models \langle C, \leq_C \rangle \text{ is a Boolean algebra.}$$

Let A be a *core* for C (see Chapter 1). We shall see that A carries the structure of a Boolean algebra in a natural way.

First we define a relation $\leq_A$ on A by putting, for each $a, a' \in A$,

$$a \leq_A a' \leftrightarrow [\![a \leq_C a']\!] = 1$$

(*cf.* proof of Lemma 1.43). It is easily verified that $\leq_A$ is a partial ordering on A and that with this partial ordering A is a Boolean algebra in which the Boolean operations $\wedge_A, \vee_A, *_A$ are given by

$$a \wedge_A a' = \text{unique } x \in A \text{ for which } [\![x = a \wedge_C a']\!] = 1;$$
$$a \vee_A a' = \text{unique } x \in A \text{ for which } [\![x = a \vee_C a']\!] = 1;$$
$$a^{*_A} = \text{unique } x \in A \text{ for which } [\![x = a^{*_C}]\!] = 1,$$

where $\wedge_C, \vee_C, *_C$ are the Boolean operations in C (in $V^{(B)}$). It is also not hard to see that, despite the apparent freedom in the choice of the core A, the Boolean algebra $\langle A, \leq_A \rangle$ is determined uniquely up to isomorphism. *We shall write $B \otimes C$ for A and $\leq_{B \otimes C}$ or $\leq$ for $\leq_A$.*

If $x, y \in B \otimes C$ and $b \in B$, then the two-term mixture $b \cdot x + b^* \cdot y$ is, with probability 1, an element of C, and hence there is a unique element of $B \otimes C$ which is equal to it with probability 1. Without loss of generality we may and shall assume in the sequel that this element of $B \otimes C$ is $b \cdot x + b^* \cdot y$. That is, *we shall assume that $B \otimes C$ is closed under two-term mixtures of the form $b \cdot x + b^* \cdot y$.* This fact should be borne in mind when considering the next problem.

Problem 6.10 (An isomorphism of Boolean algebras) Show that the map $p : B \to B \otimes \hat{2}$ defined by $p(b) = b.\hat{1} + b^*.\hat{0}$ an isomorphism of Boolean algebras.

Next, we treat the computation of arbitrary joins in $B \otimes C$.

Lemma 6.11 *Let $X \subseteq B \otimes C$ and put $X' = X \times \{1\}$. Then $X' \in V^{(B)}$ and $[\![X' \subseteq C]\!] = 1$. If $a \in B \otimes C$ satisfies $[\![a = \bigvee_C X']\!] = 1$, then $a = \bigvee X$ in $B \otimes C$.*

Proof If $x \in X$, then clearly $[\![x \in X']\!] = 1$, so that $[\![x \leq_C a]\!] = 1$, whence $x \leq_{B \otimes C} a$. Therefore a is an upper bound for X in $B \otimes C$. Also, if $y \in B \otimes C$ is any upper bound for X, then $[\![x \leq_C y]\!] = 1$ for all $x \in X$, whence $[\![y \text{ is an upper bound for } X']\!] = 1$, so that $[\![a \leq_C y]\!] = 1$, and therefore $a \leq_{B \otimes C} y$. So $a = \bigvee X$ in $B \otimes C$ as claimed. $\qquad\square$

As an immediate consequence we have the

Corollary 6.12 *If*

$$V^{(B)} \models \langle C, \leq_C \rangle \text{ is a complete Boolean algebra,}$$

then $B \otimes C$ is complete.

Next, we show that B is completely embeddable in $B \otimes C$. In $V^{(B)}$ we have the natural monomorphism i of the two element Boolean algebra $\hat{2}$ into C which sends $\hat{0}$ to 0_C and $\hat{1}$ to 1_C, where $0_C, 1_C$ are the unique elements of $B \otimes C$ which with probability 1 are the bottom and top elements of C respectively. This in turn induces the natural map $j : B \otimes \hat{2} \to B \otimes C$ defined by setting

$$j(x) = \text{ unique } y \in B \otimes C \text{ for which } [\![y = i(x)]\!] = 1.$$

Clearly, for $x \in B \otimes \hat{2}$,

$$[\![x = \hat{1}]\!] = [\![j(x) = 1_C]\!];$$
$$[\![x = \hat{0}]\!] = [\![j(x) = 0_C]\!],$$

and so $j(x)$ can be described as the two-term mixture

$$j(x) = [\![x = \hat{1}]\!] \cdot 1_C + [\![x = \hat{0}]\!] \cdot 0_C$$

for $x \in B \otimes \hat{2}$. By Problem 6.10, we have a natural isomorphism $p : B \cong B \otimes \hat{2}$ given by $p(b) = b \cdot \hat{1} + b^* \cdot \hat{0}$ for $b \in B$. Thus the composite $e = j \circ p$ is given by

$$e(b) = j(p(b)) = [\![p(b) = \hat{1}]\!] \cdot 1_C + [\![p(b) = \hat{0}]\!] \cdot 0_C = b \cdot 1_C + b^* \cdot 0_C.$$

The map $e : B \to B \otimes C$ is then given by

$$\text{for } b \in B, e(b) = \text{the unique } x \in B \otimes C \text{ for which}$$
$$[\![x = 1_C]\!] = b, [\![x = 0_C]\!] = b^*. \tag{6.13}$$

Observe that by the very definition of e we have for $b \in B$

$$V^{(B)} \models e(b) \in \{0_C, 1_C\}. \tag{6.14}$$

Recall that a homomorphism h of Boolean algebras is said to be *complete* if it preserves arbitrary joins, that is, if whenever X is a subset of the domain algebra of h such that $\bigvee X$ exists, then $\bigvee h[X]$ exists in the codomain algebra of h and is equal to $h(\bigvee X)$.

Lemma 6.15 *The map e is a complete monomorphism of B into $B \otimes C$.*

Proof The fact that e is injective and preserves complements is easily established and is left to the reader. We show that e preserves arbitrary joins in B. Suppose then that $X \subseteq B$. Let $Y = e[X]$ and $Y' = Y \times \{1\}$. Then $[\![Y' \subseteq C]\!] = 1$ and, if we choose $a \in B \otimes C$ to satisfy $[\![a = \bigvee_C Y']\!] = 1$, then by Lemma 6.11 we have $a = \bigvee Y$ in $B \otimes C$. Now (6.14) gives $[\![Y' \subseteq \{0_C, 1_C\}]\!] = 1$, so that $[\![a \in \{0_C, 1_C\}]\!] = 1$, whence

$$[\![a = 1_C]\!] = \left[\!\!\left[\bigvee_C Y' = 1_C \wedge Y' \subseteq \{0_C, 1_C\} \right]\!\!\right]$$

$$= [\![1_C \in Y']\!]$$

$$= \bigvee_{x \in X} [\![e(x) = 1_C]\!]$$

$$= \bigvee X.$$

Therefore a satisfies the defining equations (6.13) for $e(\bigvee X)$ in $B \otimes C$; in other words $e(\bigvee X) = \bigvee e[X]$. Thus e is complete as claimed. $\qquad\square$

In view of Lemma 6.15, e is called the *canonical embedding* of B in $B \otimes C$. We continue with some technical lemmas which we shall require later on.

Lemma 6.16 *For $x, y \in B \otimes C, b \in B$, we have*

$$b \le [\![x \le_C y]\!] \leftrightarrow e(b) \wedge x \le_{B \otimes C} y.$$

Proof We have

$$b = [\![e(b) = 1_C]\!].$$

So

$$b \le [\![x \le_C y]\!] \leftrightarrow V^{(B)} \models [\![e(b) = 1_C \rightarrow x \le_C y]\!]. \tag{$*$}$$

But by (6.14)

$$V^{(B)} \models e(b) = 1_C \vee e(b) = 0_C.$$

So

$$V^{(B)} \models e(b) = 1_C \to x \leq_C y \leftrightarrow V^{(B)} \models e(b) \wedge x \leq_C y$$
$$\leftrightarrow e(b) \wedge x \leq_{B \otimes C} y.$$

The result now follows from $(*)$. $\qquad\square$

Lemma 6.17 *Let* $X \in V^{(B)}$. *Then there is* $Y \in V^{(B)}$ *such that* $\mathrm{dom}(Y) \subseteq B \otimes C, Y$ *is definite* (1.40) *and*

$$[\![X \subseteq C]\!] \leq [\![Y = X \cup \{0_C\}]\!].$$

Proof Put $b = [\![X \subseteq C]\!]$. Using the Mixing Lemma, choose $X' \in V^{(B)}$ to satisfy

$$[\![X' = X \cup \{0_C\}]\!] \geq b,$$
$$[\![X' = \{0_C\}]\!] \geq b^*.$$

Then $[\![\emptyset \neq X' \subseteq C]\!] = 1$. Now put

$$Y' = \{ y \in B \otimes C : [\![y \in X']\!] = 1 \}$$

and $Y = Y' \times \{1\}$. Notice that Y' is a core for X'.

We claim that Y meets the required conditions; to establish this it clearly suffices to show that $[\![Y = X']\!] = 1$.

First, we have, for any $x \in V^{(B)}$,

$$[\![x \in Y]\!] = \bigvee_{y \in Y'} [\![x = y]\!]$$
$$= \bigvee_{y \in Y'} [\![x = y]\!] \wedge [\![y \in X']\!]$$
$$\leq [\![x \in X']\!],$$

so $[\![Y \subseteq X']\!] = 1$. Moreover, since $[\![X' \neq \emptyset]\!] = 1$, by Lemma 1.32 there is, for each $x \in V^{(B)}$, an $x' \in Y'$ such that $[\![x = x']\!] = [\![x \in X']\!]$. Hence

$$\begin{aligned}
[\![x \in X']\!] &= [\![x = x']\!] \\
&\leq \bigvee_{y \in Y'} [\![x = y]\!] \\
&= [\![x \in Y]\!],
\end{aligned}$$

so $[\![X' \subseteq Y]\!] = 1$. Thus the claim, and the lemma, are proved. $\qquad\square$

We recall from Problem 3.12(i) that the complete monomorphism $e : B \to B \otimes C$ induces a natural map $\bar{e} : V^{(B)} \to V^{(B \otimes C)}$. We shall *identify B with its image in $B \otimes C$, so that B becomes identified as a complete subalgebra of $B \otimes C$ and $V^{(B)}$ as a subclass of $V^{(B \otimes C)}$.* (Notice that this amounts to taking e, and hence $\bar{e}$, as the identity map.)

Since

$$V^{(B)} \models C \text{ is a } P^{(B)}(C)\text{-complete Boolean algebra,}$$

it follows from Corollary 1.21 that

$$V^{(B \otimes C)} \models C \text{ is a } P^{(B)}(C)\text{-complete Boolean algebra.}$$

We are going to show that, in $V^{(B \otimes C)}$, C contains a $P^{(B)}(C)$-complete ultrafilter.

Define the object $U^+ \in V^{(B \otimes C)}$ by $\mathrm{dom}(U^+) = B \otimes C$ and $U^+(a) = a$ for all $a \in B \otimes C$. That is, U^+ is the identify map on $B \otimes C$. Then, for $a \in B \otimes C$, we have

$$[\![a \in U^+]\!] = a. \tag{6.18}$$

To prove this, we note first that since the map e is now the identify, Lemma 6.16 becomes, for $x, y \in B \otimes C$, $b \in B$,

$$b \leq [\![x \leq_C y]\!] \leftrightarrow b \wedge x \leq_{B \otimes C} y. \tag{6.19}$$

Now $[\![x = a]\!] \leq [\![x \leq_C a]\!]$, so taking $b = [\![x = a]\!]$ we get from Lemma 6.16,

$$[\![x = a]\!] \wedge x \leq_{B \otimes C} a.$$

Thus

$$[\![a \in U^+]\!] = \bigvee_{x \in B \otimes C} [\![x = a]\!] \wedge x \le a$$

and the reverse inequality follows from Theorem 1.17 (ii).

Our next result is crucial.

Theorem 6.20

$$V^{(B \otimes C)} \models U^+ \text{ is a } P^{(B)}(C)\text{-complete ultrafilter in } C.$$

Proof The proof proceeds somewhat along the lines of Theorem 4.21 (of which the present theorem is actually a generalization), only it is a little more troublesome. We shall only verify two of the properties that U^+ must have, leaving the verification of the others to the reader.

(a) $[\![\forall xy \in U^+[x \wedge_C y \in U^+]]\!] = 1$. To verify this, observe that the l.h.s. is:

$$\bigwedge_{a,b \in B \otimes C} [U^+(a) \wedge U^+(b)] \Rightarrow [\![a \wedge_C b \in U^+]\!] = \bigwedge_{a,b \in B \otimes C} [a \wedge b \Rightarrow a \wedge b] = 1.$$

(b) $[\![\forall X \in P^{(B)}(C)[\bigvee_C X \in U^+ \rightarrow U^+ \cap X \neq \emptyset]]\!] = 1$. To verify this, notice that the l.h.s. is

$$\bigwedge_{X \subseteq \mathrm{dom}(P^{(B)}(C))} \left[\!\!\left[X \subseteq C \wedge \bigvee_C X \in U^+ \right]\!\!\right] \Rightarrow [\![U^+ \cap X \neq \emptyset]\!].$$

Given $X \in \mathrm{dom}(P^{(B)}(C))$, take $Y \in V^{(B)}$ to satisfy the conditions of Lemma 6.17. Then

$$\left[\!\!\left[X \subseteq C \wedge \bigvee_C X \in U^+ \right]\!\!\right] \le \left[\!\!\left[\bigvee_C Y \in U^+ \right]\!\!\right].$$

Now, by Lemma 6.11 there is an $a \in B \otimes C$ such that $[\![a = \bigvee_C Y]\!] = 1$ and also such that $a = \bigvee_{B \otimes C} \mathrm{dom}(Y)$. Therefore

$$\left[\!\!\left[\bigvee_C Y \in U^+ \right]\!\!\right] = [\![a \in U^+]\!] = a = \bigvee_{B \otimes C} \mathrm{dom}(Y).$$

Hence

$$\left[\!\!\left[X \subseteq C \wedge \bigvee_C X \in U^+ \right]\!\!\right] \leq \bigvee_{B \otimes C} \mathrm{dom}(Y)$$

$$= \bigvee_{x \in \mathrm{dom}(Y)} [\![x \in U^+]\!]$$

$$= [\![\exists x \in Y [x \in U^+]]\!]$$

$$= [\![U^+ \cap Y \neq \emptyset]\!].$$

So

$$\left[\!\!\left[X \subseteq C \wedge \bigvee_C X \in U^+ \right]\!\!\right] \leq [\![U^+ \cap Y \neq \emptyset]\!] \wedge [\![X \subseteq C]\!]$$

$$\leq [\![U^+ \cap Y \neq \emptyset \wedge Y = X \cup \{0_C\}]\!]$$

$$\leq [\![U^+ \cap X \neq \emptyset]\!],$$

and (b) follows. $\qquad\qquad\qquad\qquad\qquad\qquad\qquad\qquad\qquad\qquad\square$

Remarks   (1) Taking $B = 2$ and $C = \hat{A}$ in Theorem 6.20, where A is a complete Boolean algebra in V, then $2 \otimes \hat{A} \cong A$ and it is not hard to see that U^+ is (essentially) the canonical generic ultrafilter in $V^{(A)}$. So Theorem 6.20 generalizes Theorem 4.21.

    (2) Since $V^{(B)} \subseteq V^{(B \otimes C)}$, we may regard $V^{(B)}$ as a *class* in $V^{(B \otimes C)}$. Moreover, it is not hard to see that, within $V^{(B \otimes C)}$, $V^{(B)}$ is a transitive model of ZFC containing all the ordinals. So working inside $V^{(B \otimes C)}$ we can form the C-extension $(V^{(B)})^{(C)}$ of $V^{(B)}$. Now Theorem 6.20 may be construed as asserting that, within $V^{(B \otimes C)}$, U^+ is a $V^{(B)}$-generic ultrafilter in C. Accordingly within $V^{(B \otimes C)}$ we can form the transitive collapse $V^{(B)}[U^+]$ of the quotient $(V^{(B)})^{(C)}/U^+$ and by applying Theorem 4.22 within $V^{(B \otimes C)}$ we have

$$V^{(B \otimes C)} \models V^{(B)}[U^+] \text{ is the model of ZFC generated by } U^+ \text{ and } V^{(B)}.$$

Now, writing U_* for the canonical generic ultrafilter in $V^{(B \otimes C)}$, it is shown in Problem 6.36 that

$$V^{(B \otimes C)} \models U_* \in V^{(B)}[U^+]. \tag{6.21}$$

Moreover, we know from the remarks following Lemma 4.25 that

$$V^{(B \otimes C)} \models \forall x [x \in \hat{V}[U_*]] \tag{6.22}$$

and so we get, using (6.21)

$$V^{(B \otimes C)} \models \forall x [x \in V^{(B)}[U^+]]. \tag{6.23}$$

In other words, $V^{(B \otimes C)}$ may be regarded as the Boolean-valued model of ZFC generated by $V^{(B)}$ and U^+.

Notice also that if $U_\circ$ is the canonical generic ultrafilter in $V^{(B)}$ then $V^{(B)} \models \forall x [x \in \hat{V}[U_\circ]]$ and so we get, using (6.22) and (6.23),

$$V^{(B \otimes C)} \models \forall x [x \in \hat{V}[U_\circ][U^+]]$$

$$V^{(B \otimes C)} \models \hat{V}[U_*] = \hat{V}[U_\circ][U^+].$$

The first of these expressions tells us that $V^{(B \otimes C)}$ may be regarded as a 'double generic extension' of $\hat{V}$, and the second that this double extension is expressible as a single extension.

(3) Working within $V^{(B \otimes C)}$, let $i_{U^+} = i$ be canonical map of $(V^{(B)})^{(C)}$ onto $V^{(B)}[U+]$. It follows from (6.23) that within $V^{(B \otimes C)}$ this map carries $(V^{(B)})^{(C)}$ onto the universe, that is,

$$V^{(B \otimes C)} \models \forall x [x \in i[(V^{(B)})^{(C)}]]. \tag{6.24}$$

Now let $J^{(C)}$ be the class

$$J^{(C)} = \{x \in V^{(B)} : [\![x \in V^{(C)}]\!]^B = 1\};$$

that is, $J^{(C)}$ is the class of all B-valued sets which, with probability 1, are members of $(V^{(B)})^{(C)}$. $J^{(C)}$ may be deemed to be the class that represents $(V^{(B)})^{(C)}$ in the real world. We may regard $J^{(C)}$ as a $(B \otimes C)$-valued structure by defining, for $x, y \in J^{(C)}$,

$$[\![x = y]\!]^{J^{(C)}} = \text{unique } a \in B \otimes C \text{ such that } V^{(B)} \models a = [\![x = y]\!]^C$$

$$[\![x \in y]\!]^{J^{(C)}} = \text{unique } a \in B \otimes C \text{ such that } V^{(B)} \models a = [\![x \in y]\!]^C.$$

Next, let us agree to *identify* elements of $J^{(C)}$ (and $V^{(B \otimes C)}$) when they are equal with probability 1. Having done this, we can define the map $j : J^{(C)} \to V^{(B \otimes C)}$ by putting, for each $x \in J^{(C)}$,

$$j(x) = \text{unique } y \in V^{(B \otimes C)} \text{such that } V^{(B \otimes C)} \models y = i(x).$$

Using (6.23) and (6.18) it is now easy to verify that j is an isomorphism of the Boolean-valued structures $J^{(C)}$ and $V^{B \otimes C}$. That is, j is a map of $J^{(C)}$

onto $V^{(B\otimes C)}$ such that $[\![x = y]\!]^{J^{(C)}} = [\![j(x) = j(y)]\!]^{B\otimes C}$ and $[\![x \in y]\!]^{J^{(C)}} = [\![j(x) \in j(y)]\!]^{B\otimes C}$ for all $x, y \in J^{(C)}$. Accordingly, we have shown that the 'iterated' Boolean extension $V^{(B)(C)}$ is equivalent to the 'ordinary' Boolean extension $V^{(B\otimes C)}$.

Our final result in this section is the following.

Lemma 6.25 *If B satisfies* ccc *and* $V^{(B)} \models C$ *satisfies* ccc, *then* $B \otimes C$ *satisfies* ccc.

Proof Let $A = \{a_\xi : \xi < \omega_1\}$ be an antichain in $B \otimes C$; we show that there is $\xi_0 < \omega_1$ such that $a_\eta = 0$ for all $\eta > \xi_0$. Let $A' = A \times \{1\}$; then clearly

(1) $V^{(B)} \models A'$ *is an antichain in* C.

Since C satisfies ccc in $V^{(B)}$, it follows from (1) that

(2) $V^{(B)} \models A'$ *is countable.*

Now define $f \in V^{(B)}$ by

$$f = \{\langle \hat{\xi}, a_\xi \rangle^{(B)} : \xi < \omega_1\} \times \{1\}.$$

It is then easily verified that

(3) $V^{(B)} \models f : \hat{\omega}_1 \to A'$

and, for all $\xi < \omega_1$

(4) $[\![f(\hat{\xi}) = a_\xi]\!] = 1.$

Since B satisfies ccc, we have, by Theorem 1.51,

$$V^{(B)} \models \hat{\omega}_1 \text{ is uncountable}$$

and it follows from (2) and (3) that

$$V^{(B)} \models \exists \xi < \omega_1 \forall \eta > \xi \forall \eta' > \xi [f(\eta) = f(\eta')].$$

Therefore, by the Maximum Principle, there is $\xi \in V^{(B)}$ such that

(5) $V^B \models \xi < \hat{\omega}_1 \wedge \forall \eta > \xi \forall \eta' > \xi [f(\eta) = f(n')].$

Since B satisfies ccc, by Theorem 1.51(v) there is an ordinal $\xi_0 < \omega_1$ such that $[\![\xi < \hat{\xi}_0]\!] = 1$. It follows from (5) that

$$V^{(B)} \models \forall \eta > \hat{\xi}_0 [f(\eta) = f(\hat{\xi}_0)].$$

Hence $[\![f(\hat{\eta}) = f(\hat{\xi}_0)]\!] = 1$ whenever $\xi_0 < \eta < \omega_1$. Thus, by (4), $[\![a_\xi = a_\eta]\!] = 1$ whenever $\xi_0 < \eta < \omega_1$. But since A is an antichain we have, for $\eta > \xi_0$,

$$1 = [\![a_{\xi_0} = a_\eta]\!] \leq [\![a_\eta = a_{\xi_0} \wedge a_\eta]\!] = [\![a_\eta = 0]\!].$$

Therefore $a_\eta = 0$ in $B \otimes C$ for $\eta > \xi_0$, as claimed. $\square$

Further results on Boolean algebras

In this section we give some technical results on Boolean algebras that we shall require for the proof of relative consistency of SH.

Our first result is a generalization of Lemma 2.10.

Lemma 6.26 *Let B be a complete Boolean algebra satisfying ccc, and let D be a dense subset of B. Then for each $b \in B$ there is a countable subset D_b of D such that $b = \bigvee D_b$. Moreover, $|B| \leq |D|^{\aleph_0}$.*

Proof Using Zorn's lemma, let D_b be a maximal antichain in the set $\{x \in D : x \leq b\}$. Put $a = \bigvee D_b$; we claim that $a = b$. Clearly $a \leq b$. On the other hand, consider $b - a$. If it is nonzero, then there is $d \in D$ such that $0 \neq d \leq b - a$. But then d is disjoint from every member of D_b, contradicting the latter's maximality. It follows that $b - a = 0$, so $a = b$.

Thus $b = \bigvee D_b$; and since B satisfies ccc, D_b, as an antichain, must be countable.

Consequently each member of B is determined by a countable subset of D; since there are at most $|D|^{\aleph_0}$ of these, the claimed inequality follows. $\square$

Corollary 6.27 *Suppose κ is a regular uncountable cardinal, B is a complete Boolean algebra satisfying ccc such that $|B| \leq \kappa$ and $C \in V^{(B)}$ satisfies*

$$V^{(B)} \models C \text{ is a complete Boolean algebra, satisfies}$$

$$\text{ccc and has a dense subset of cardinality} < \hat{\kappa}.$$

Then for some cardinal $\lambda < \kappa, |B \otimes C| \leq \kappa^\lambda$.

Proof Let $Q \in V^{(B)}$ be such that

$$[\![Q \text{ is dense in } C \text{ and } |Q| < \hat{\kappa}]\!] = 1$$

and let $Q' = \{x \in B \otimes C : [\![x \in Q]\!] = 1\}$. Then, using Theorem 1.51(v), there is an ordinal $\alpha < \kappa$ such that $[\![|Q| < \hat{\alpha}]\!] = 1$. Putting $\lambda = |\alpha| < \kappa$ (or $\lambda = \omega$ if α is finite), we claim first that $|Q'| \leq \kappa^\lambda$. To see this, observe that since $[\![|Q| < \hat{\alpha}]\!] = 1$ the Maximum Principle yields an $f \in V^{(B)}$ such that $[\![f \text{ is a map}$

of $\hat{\alpha}$ *onto* $Q] = 1$. Each $x \in Q'$ is then uniquely determined by the function $g_x : \alpha \to B$ defined by $g_x(\gamma) = [f(\hat{\gamma}) = x]$. Since there are at most κ^λ such functions g_x, the claim follows.

We next claim that the set

$$S = \{b \wedge x : 0 \neq b \in B \wedge x \in Q'\}$$

is dense in $B \otimes C$. For suppose $0 \neq c \in B \otimes C$. Then $[c \in C] = 1$ and $[c \neq 0_C] = b \neq 0$. Put $d = b \cdot c + b^* \cdot 1$. Then $[d = c] \geq b$, $[d \in C] = 1$ and $[d \neq 0_C] = 1$. Since $[Q$ *is dense in* $C] = 1$, there is $x \in Q'$ such that $[x \leq_C d] = 1$. It follows that

$$b \leq [c = d] = [c = d] \wedge [x \leq_C d] \leq [x \leq_C c].$$

So, by (6.19), $b \wedge x \leq c$, and the claim follows.

Now we have $|S| \leq |Q'| \cdot |B| \leq \kappa^\lambda \cdot \kappa = \kappa^\lambda$. Therefore, by Lemma 6.26,

$$|B \otimes C| \leq |S|^{\aleph_0} \leq \kappa^{\lambda \cdot \aleph_0} = \kappa^\lambda.$$

$$\square$$

We recall (see the remark after Lemma 2.3) that for each Boolean algebra A there is a complete Boolean algebra B called the *completion* of A such that (if we identify A as a subalgebra of B via the canonical monomorphism $f : A \to B$)

(i) A is a subalgebra of B;

(ii) $A - \{0\}$ is dense in B;

(iii) if $X \subseteq A$ has a join $\bigvee_A X$ in A, then $\bigvee_A X = \bigvee_B X$.

Let $\langle B_i : i \in 1 \rangle$ be a chain of (complete) Boolean algebras; that is, such that for any pair B_i, B_j, one is a subalgebra of the other. Then we can form the direct limit $\varinjlim B_i$ of the chain in the category of Boolean algebras in the usual way: $\varinjlim B_i$ is just $\bigcup_{i \in I} B_i$ with Boolean operations inherited from the B_i in the obvious manner. The completion of $\varinjlim B_i$ is called the *limit* completion of the chain $\langle B_i : i \in I \rangle$ and is written $\mathrm{limco}_{i \in I} B_i$ or $\mathrm{limco} B_i$. Clearly $\mathrm{limco} B_i$ includes $\bigcup_{i \in I} B_i - \{0\}$ as a dense subset.

Now let α be a limit ordinal. A sequence $\langle B_\xi : \xi < \alpha \rangle$ of complete Boolean algebras is called a *normal* sequence if

(a) $B_0 = 2$;

(b) for $\xi < \eta$, B_ξ is a complete subalgebra of B_η;

(c) for limit β, $B_\beta = \mathrm{limco}_{\xi < \beta} B_\xi$.

Lemma 6.28 *Let $\langle B_\xi : \xi < \alpha \rangle$ be a normal sequence of complete Boolean algebras and let $B = \mathrm{limco}_{\xi<\alpha} B_\xi$. Then each B_ξ is a complete subalgebra of B. Suppose, further, that each B_ξ satisfies ccc and that α is an uncountable regular cardinal. Then*

(i) $B = \bigcup_{\xi<\alpha} B_\xi$;

(ii) *if $|X| < \alpha$ and $f : X \to B$, then $\mathrm{ran}(f) \subseteq B_\xi$ for some $\xi < \alpha$.*

Proof Let $X \subseteq B_\xi$ and $a = \bigvee X$ in B_ξ. We claim that $a = \bigvee X$ in $\varinjlim B_\eta = B'$. For if not, then X would have an upper bound $b < a$ in B'. Since $B' = \bigcup B_\eta$ there is $\eta < \alpha$ such that $b \in B_\eta$, and without loss of generality we may suppose that $\xi \leq \eta$. But then, in B_η, $\bigvee X \leq b < a$, contradicting the assumption that B_ξ is a complete subalgebra of B_η. Therefore $\bigvee_B X = \bigvee_{B'} X$ and, since B is the completion of B', $\bigvee_{B'} X = \bigvee_B X$. So B_ξ is complete subalgebra of B.

We next prove (i). Let $x \in B$. Since $\bigcup_{\xi<\alpha} B_\xi - \{0\}$ is dense in B, by Lemma 6.26 there is a countable set $\{x_n : n \in \omega\}$ such that $x_n \in B_{\xi_n}$ with $\xi_n < \alpha$ and $x = \bigvee_{n\in\omega} x_n$. Since λ is regular, there is $\xi < \alpha$ such that $\xi \geq \xi_n$ for all n, so that $x_n, \in B_\xi$ for all n, whence $x = \bigvee_{n\in\omega} x_n \in B_\xi$. (i) follows.

Finally, (ii) follows easily from (i) and the regularity of α. $\square$

Corollary 6.29 *Let κ be an uncountable regular cardinal, $\langle B_\xi : \xi < k \rangle$ be a normal sequence of complete Boolean algebras satisfying ccc and let $B = \mathrm{limco}B_\xi$. Suppose further that $X \in V^{(B)}$ satisfies (a) $[\![|X| < \kappa]\!] = 1$ and (b) either $[\![X \subseteq \hat\kappa]\!] = 1$ or $[\![X \subseteq \hat\kappa \times \hat\kappa]\!] = 1$ or $[\![X \subseteq P\hat\kappa \wedge |\bigcup X| < \hat\kappa]\!] = 1$. Then there are $\xi < \kappa$ and $Y \in V^{(B_\xi)}$ such that $[\![Y = X]\!] = 1$.*

Proof Since κ and $\kappa \times \kappa$ are naturally bijective and each $X \subseteq P\kappa$ such that $|X| < \kappa$ is naturally correlated with a subset of $\kappa \times \kappa$, the proof reduces to the case in which $[\![X \subseteq \hat\kappa]\!] = 1$. By Theorem 1.51(iv) we know that $[\![\hat\kappa \text{ is regular}]\!] = 1$ and since $[\![|X| < \hat\kappa]\!] = 1$ it follows (using the Maximum Principle) that there is $\alpha \in V^{(B)}$ such that

$$[\![\alpha < \hat\kappa \wedge \forall \xi \geq \alpha[\xi \notin X]]\!] = 1.$$

By Theorem 1.51(v) there is an ordinal $\gamma < \kappa$ such that $[\![\alpha < \hat\gamma]\!] = 1$. It follows that

(1) $$[\![\hat\xi \in X]\!] = 0 \text{ for all } \xi \geq \gamma.$$

Now define $Y \in V^{(B)}$ by $\mathrm{dom}(Y) = \{\hat\xi : \xi < \gamma\}$ and $Y(\hat\xi) = [\![\hat\xi \in X]\!]$. It is then easily verified, using (1), that $[\![X = Y]\!] = 1$. So it remains to show that

$Y \in V^{(B_\xi)}$ for some $\xi < \kappa$. But this follows immediately from an application of Lemma 6.28 to the map $f : \gamma \to B$ given by $f(\xi) = Y(\xi)$. $\qquad\square$

Now let B be a complete Boolean algebra and let C be a complete subalgebra of B. We define the map $\pi : B \to C$ as follows.

$$\pi(x) = \bigwedge \{y \in C : x \leq y\}.$$

Notice that we then have, for $b \in B, c \in C$, the '*adjointness*' condition:

$$b \leq c \leftrightarrow \pi(b) \leq c, \tag{6.30}$$

which in turn characterizes π uniquely. The π is called the *canonical projection of B onto C.*

Lemma 6.31 *The canonical projection $\pi : B \to C$ has the following properties:*

    (i) $\pi(x) \geq x$;
    (ii) $x \leq y \to \pi(x) \leq \pi(y)$;
    (iii) $\pi(c) = c$ *for all $c \in C$;*
    (iv) $\pi(\bigvee X) = \bigvee \pi[X]$ *for all $X \subseteq B$;*
    (v) $\pi(b) \wedge c = \pi(b \wedge c)$ *for all $b \in B, c \in C$.*

Proof (i), (ii), and (iii) are obvious. To prove (iv), we note that, if $c \in C$ then

$$\pi(\bigvee X) \leq c \leftrightarrow \bigvee X \leq c \quad \text{(by (6.30))}$$
$$\leftrightarrow \forall x \in X[x \leq c]$$
$$\leftrightarrow \forall x \in X[\pi(x) \leq c] \quad \text{(by (6.30))}$$
$$\leftrightarrow \bigvee \pi[X] \leq c.$$

As for (v), we have $\pi(b) \wedge c = \pi(b) \wedge \pi(c) \geq \pi(b \wedge c)$ by (ii) and (iii). On the other hand,

$$\pi(b) = \pi(b \wedge c) \vee \pi(b \wedge c^*) \quad \text{(by (iv))}$$
$$\leq \pi(b \wedge c) \vee c^* \quad \text{(by (ii) and (iii)).}$$

So $\pi(b) \wedge c \leq \pi(b \wedge c) \wedge c \leq \pi(b \wedge c)$ and the result follows. $\qquad\square$

Our final preparatory result concerns the preservation of the countable chain condition under passage to limit completions.

Theorem 6.32 *Let $\langle B_\xi : \xi < \alpha \rangle$ be a normal sequence of complete Boolean algebras, and suppose that each B_ξ satisfies* ccc. *Then the limit completion B of $\langle B_\xi : \xi < \alpha \rangle$ also satisfies* ccc.

Proof We first reduce the proof of the theorem to the case in which $\alpha = \omega_1$. Suppose the hypothesis of the theorem true and the conclusion false. Let A be an antichain in B of cardinality $\aleph_1$ such that $0 \notin A$. Since $\bigcup_{\xi < \alpha} B_\xi$ is dense in B, for each $a \in A$ we can find an ordinal $\gamma_a < \alpha$ and an element $b_a \in B_{\gamma_a}$ such that $0 \neq b_a \leq a$. Clearly $\{b_a : a \in A\}$ is an antichain in B of cardinality $\aleph_1$. Also, sup $\{\gamma_a : a \in A\} = \alpha$, for otherwise there would be an ordinal $\beta < \alpha$ for which $\gamma_a < \beta$ for $a \in A$, contradicting the assumption that B_β satisfies ccc. Since $|A| = \aleph_1$, it follows that α is cofinal with ω_1.

On the other hand, α is not cofinal with ω. For if $\alpha = \sup\{\alpha_n : n \in \omega\}$ then for each $a \in A$ there is $n \in \omega$ such that $\gamma_a < \alpha_n$. Hence $\{b_a : a \in A\} \subseteq \bigcup_{n \in \omega} B_{\alpha_n}$ and so, for some n, $B_{\alpha_n} \cap \{b_a : a \in A\}$ must have cardinality $\aleph_1$. This contradicts the assumption that B_{α_n} satisfies ccc.

It follows that α has cofinality ω_1. Accordingly there is a sequence of ordinals $\langle \beta_\xi : \xi < \omega_1 \rangle$ such that

$$\beta_0 = 0, \eta < \xi \to \beta_\eta < \beta_\xi, \alpha = \sup\{\beta_\xi : \xi < \omega_1\}$$

$$\text{and} \quad \beta_\xi = \sup\{\beta_\eta : \eta < \xi\} \text{ for limit } \xi.$$

If we put B'_ξ for B_{β_ξ}, then $\langle B'_\xi : \xi < \omega_1 \rangle$ is a normal sequence, each B_ξ satisfies ccc, but the limit completion B_α of $\langle B'_\xi : \xi < \omega_1 \rangle$ does not.

Thus we need only prove the theorem when $\alpha = \omega_1$. Let X be an antichain in B; we show that X is countable.

Without loss of generality we may assume that $\bigvee X = 1$ and, since $\bigcup_{\xi < \omega_1} B_\xi - \{0\}$ is dense in B, that $X \subseteq \bigcup_{\xi < \omega_1} B_\xi$. For $\xi < \omega_1$, put $X_\xi = X \cap B_\xi$. Since B_ξ satisfies ccc, X_ξ is countable.

Now by Lemma 6.28, B_ξ is a complete subalgebra of B and so we can consider the canonical projection π_ξ of B onto B_ξ. By Lemma 6.31 (iv), we have

$$1 = \pi_\xi \left(\bigvee X \right) = \bigvee \pi_\xi[X].$$

Since B_ξ satisfies ccc, there is, by Problem 1.53 (iv), a countable subset X' of X such that

$$1 = \bigvee \pi_\xi[X'].$$

It follows that for some countable ordinal δ_ξ we have $X' \subseteq X_{\delta_\xi}$ and

$$1 = \bigvee \pi_\xi[X_{\delta_\xi}]. \tag{1}$$

Now let γ be a countable limit ordinal such that $\delta_\xi < \gamma$ for all $\xi < \gamma$. We claim that $X = X_\gamma$. Since X_γ is countable, this will complete the proof.

For $\xi < \gamma$ we have

$$\bigvee \pi_\xi[X_{\delta_\xi}] = \pi_\xi\left(\bigvee X_{\delta_\xi}\right) \leq \pi_\xi\left(\bigvee X_\gamma\right),$$

so it follows from (1) that

$$1 = \pi_\xi\left(\bigvee X_\gamma\right). \tag{2}$$

We claim that $\bigvee X_\gamma = 1$. For if not, then since $\bigcup_{\xi<\gamma} B_\xi - \{0\}$ is dense in B_γ, there is $\xi < \gamma$ and $0 \neq b \in B_\xi$ such that $b \wedge \bigvee X_\gamma = 0$. Hence, by Lemma 6.31(v)

$$0 = \pi_\xi\left(b \wedge \bigvee X_\gamma\right) = b \wedge \pi_\xi\left(\bigvee X_\gamma\right).$$

But this contradicts (2).

Therefore $\bigvee X_\gamma = 1$ and it follows easily from this that $X = X_\gamma$. $\square$

Notice that Theorem 6.32 yields as an immediate corollary the result that the direct limit of a normal sequence of complete Boolean algebras satisfying ccc also satisfies ccc.

The relative consistency of SH

At long last we are in a position to prove the relative consistency of $\mathrm{MA} + 2^{\aleph_0} > \aleph_1$, and so of SH.

Theorem 6.33 *Let κ be an uncountable regular cardinal such that for any $0 \neq \lambda < \kappa$ we have $\kappa^\lambda = \kappa$. Then there is a complete Boolean algebra B such that*

$$V^{(B)} \models \mathrm{MA} + 2^{\aleph_0} = \hat{\kappa}.$$

Proof We construct recursively a normal sequence $\langle B_\xi : \xi < \kappa \rangle$ of complete Boolean algebras such that, for all $\xi < \kappa$, (a) B_ξ satisfies ccc and (b) $|B_\xi| \leq \kappa$.

For each complete Boolean algebra B satisfying ccc such that $|B| \leq \kappa$, let $D(B)$ be a core for the $V^{(B)}$-set $\{x : x \subseteq \hat{\kappa} \times \hat{\kappa} \wedge |x| < \hat{\kappa}\}^{(B)}$. We claim that $|D(B)| \leq \kappa$. To prove this we note first that if $x \in D(B)$ then, since $[\![\hat{\kappa}\ is\ regular\ ]\!] = 1$ by Theorem 1.51(iv) there is $\alpha \in V^{(B)}$ such that $[\![\alpha < \hat{\kappa} \wedge x \subseteq \alpha \times \alpha]\!] = 1$ and hence by Theorem 1.51(v) an ordinal $\beta < \kappa$ such that $[\![\alpha < \hat{\beta}]\!] = 1$. Therefore $[\![x \subseteq \hat{\beta} \times \hat{\beta}]\!] = 1$. It follows that if for each $\beta < \kappa$ we let v_β be a core for the $V^{(B)}$-set $P^{(B)}(\hat{\beta} \times \hat{\beta})$ then $|D(B)| \leq |\bigcup_{\beta<\kappa} v_\beta| \leq \Sigma_{\beta<\kappa}|v_\beta|$. Now if for each $z \in v_\beta$ we define $f_z : \beta \times \beta \to B$ by $f_z(\xi, \eta) = [\![\langle \hat{\xi}, \hat{\eta} \rangle \in z]\!]$, then it is easily verified that the map $z \mapsto f_z$ is a bijection between v_β and $B^{\beta \times \beta}$. It follows that $|v_\beta| \leq \kappa^{|\beta|} = \kappa$. Therefore $|D(B)| \leq \Sigma_{\beta < \kappa} |v_\beta| \leq \kappa \cdot \kappa = \kappa$ as claimed.

For each complete Boolean algebra B satisfying ccc such that $|B| \leq \kappa$, we fix an enumeration $\langle R^B_\xi : \xi < \kappa \rangle$ of $D(B)$.

Now we put $B_0 = 2$, and for α satisfying $0 < \alpha < \kappa$ assume as inductive hypothesis that $\langle B_\xi : \xi < \alpha \rangle$ has been constructed so as to be a normal sequence satisfying (a) and (b).

If α is a limit ordinal, we put $B_\alpha = \mathrm{limco}_{\xi < \alpha} B_\xi$. If α is a successor ordinal, say $\xi + 1$, we construct $B_\alpha = B_{\xi+1}$ as follows. Let $\xi \mapsto \langle \beta_\xi, \gamma_\xi \rangle$ be the canonical map of κ onto $\kappa \times \kappa$ (see Chapter 0). Recall that $\beta_\xi \leq \xi$ for any $\xi < \kappa$. Let $A = B_{\beta_\xi}$; then by inductive hypothesis A is a complete subalgebra of B_ξ and $|A| \leq \kappa$. Putting $R = R^A_{\gamma_\xi}$, we have $R \in V^{(B_\xi)}$. In $V^{(B_\xi)}$, let $\Delta(R) = \langle \{x : \langle x, x \rangle \in R\}, R \rangle$, and put

$$b = [\![\Delta(R) \text{ is a Boolean algebra satisfying ccc}]\!].$$

Let C'_ξ be the two-term mixture:

$$C'_\xi = b \cdot \Delta(R) + b^* \cdot \hat{2};$$

then

$$[\![C'_\xi \text{ is a Boolean algebra satisfying ccc}]\!] = 1.$$

Let $C_\xi \in V^{(B)}$ satisfy

$$V^{(B_\xi)} \models C_\xi \text{ is the completion of } C'_\xi.$$

Note that

$$V^{(B_\xi)} \models C_\xi \text{ satisfies ccc.}$$

We finally put

$$B_{\xi+1} = B_\xi \otimes C_\xi,$$

and identify B_ξ with its canonical image in $B_\xi \otimes C_\xi$, so that B_ξ becomes a complete subalgebra of $B_{\xi+1}$.

Clearly $\langle B_\xi : \xi < \kappa \rangle$ constructed in this way is a normal sequence. We now verify (a) and (b).

(a) is proved by induction on ξ. $B_0 = 2$ clearly satisfies ccc. If B_ξ satisfies ccc, then $B_{\xi+1} = B_\xi \otimes C_\xi$ satisfies ccc by Lemma 6.25. If α is a limit ordinal and B_ξ satisfies ccc for all $\xi < \alpha$, then B_α, as the limit completion of the normal sequence $\langle B_\xi : \xi < \alpha \rangle$, satisfies ccc by Theorem 6.32. So (a) follows.

(b) is also proved by induction on ξ. Clearly $|B_0| \leq \kappa$. If α is a limit ordinal then B_α has the dense subset $\bigcup_{\xi<\alpha} B_\xi - \{0\}$ of cardinality $\leq \kappa$ by inductive hypothesis, so, since B_α satisfies ccc, $|B_\alpha| \leq \kappa^{\aleph_0} = \kappa$ by Lemma 6.26. If $\xi < \kappa$ and $|B_\xi| \leq \kappa$, we observe that

$$V^{(B_\xi)} \models C_\xi \text{ satisfies ccc and has a dense subset } C'_\xi \text{ such that } |C'_\xi| < \hat{\kappa},$$

so $|B_{\xi+1}| = |B_\xi \otimes C_\xi| \leq \kappa$ by Corollary 6.27.

Now we define B to be the limit completion of $\langle B_\xi : \xi < \kappa \rangle$. Then by Theorem 6.32 B satisfies ccc and since it has the dense subset $\bigcup_{\xi<\kappa} B_\xi - \{0\}$ of cardinality $\leq \kappa$, it follows from Lemma 6.26 that $|B| \leq \kappa^{\aleph_0} = \kappa$. Therefore by Problem 2.19(i), we have

$$V^{(B)} \models 2^{\aleph_0} \leq (\kappa^{\aleph_0})\hat{\;} = \hat{\kappa}.$$

Thus, if we can show that $\mathrm{MA}_{\hat{\kappa}}$ holds in $V^{(B)}$, we will have, by Lemma 6.6, $V^{(B)} \models \hat{\kappa} \leq 2^{\aleph_0}$, whence $V^{(B)} \models \hat{\kappa} = 2^{\aleph_0}$ and so, in $V^{(B)}$, $\mathrm{MA}_{\hat{\kappa}}$ is just MA. It will therefore follow that

$$V^{(B)} \models 2^{\aleph_0} = \hat{\kappa} \wedge \mathrm{MA}.$$

So it remains to show that

$$V^{(B)} \models \mathrm{MA}_{\hat{\kappa}}.$$

By Theorem 6.8 and Corollary 1.28(ii) it suffices to show that, if $A, R, S \in V^{(B)}$ satisfy

(1) $V^{(B)} \models \langle A, R \rangle$ is Boolean algebra satisfying ccc $\wedge |A| < \hat{\kappa} \wedge S \subseteq PA \wedge |S| < \hat{\kappa}$,

then

$$V^{(B)} \models \text{there is an } S\text{-complete ultrafilter in } \langle A, R \rangle.$$

Without loss of generality we may assume that $V^{(B)} \models A \subseteq \hat{\kappa}$ and so

(2) $V^{(B)} \models R \subseteq \hat{\kappa} \times \hat{\kappa} \wedge |R| < \hat{\kappa} \wedge S \subseteq P\hat{\kappa} \wedge |S \cup \bigcup S| < \hat{\kappa}$.

By Corollary 6.29, there exist $\xi < \kappa$ and $A', R', S' \in V^{(B_\xi)}$ such that $[\![A = A']\!] = [\![R = R']\!] = [\![S = S']\!] = 1$ and so we may assume that $A, R, S \in V^{(B_\xi)}$.

Now we claim that

(3) $V^{(B_\xi)} \models \langle A, R \rangle$ is a Boolean algebra satisfying ccc $\wedge |A| < \hat{\kappa} \wedge S \subseteq PA \wedge |S| < \hat{\kappa}$.

Suppose, for example, $[\![|A| < \hat{\kappa}]\!]^{B_\xi} \neq 1$. Then $[\![|A| \geq \hat{\kappa}]\!]^{B_\xi} \neq 0$, so, for some $f \in V^{(B_\xi)}$, $[\![f : \hat{\kappa} \xrightarrow{\text{one--one}} A]\!]^{B_\xi} \neq 0$. By Corollary 1.21, and the fact that B_ξ is a complete subalgebra of B, $[\![f : \hat{\kappa} \xrightarrow{\text{one--one}} A]\!]^B \neq 0$. Therefore $[\![|\hat{\kappa}| \leq |A|]\!]^B \neq 0$. But since B satisfies ccc, we have $[\![\hat{\kappa} = |\hat{\kappa}|]\!]^B = 1$, so $[\![\hat{\kappa} \leq |A|]\!] \neq 0$. This contradicts (1).

Similarly, if

$$[\![\langle A, R \rangle \ \textit{does not satisfy ccc}]\!]^{B_\xi} \neq 0,$$

then there is $X \in V^{(B_\xi)}$ such that

$$[\![X \ \textit{is an antichain in } A \wedge |X| \geq \aleph_1 = \hat{\aleph}_1]\!]^{B_\xi} \neq 0.$$

So by Corollary 1.21

$$[\![X \ \textit{is an antichain in } A \wedge |X| \geq \aleph_1]\!]^B \neq 0,$$

whence

$$[\![X \ \textit{is an uncountable antichain in } A]\!]^B \neq 0.$$

But this contradicts (1). So (3) is proved.

It follows from (2) and the definition of $D(B_\xi)$ that there is $R'' \in D(B_\xi)$ such that $[\![R = R'']\!]^{B_\xi} = 1$ and so without loss of generality we may assume that $R \in D(B_\xi)$. Choose $\eta < \kappa$ to satisfy $R = R_\eta^{B_\xi}$, and $\alpha < \kappa$ so that

$$\beta_\alpha = \xi, \gamma_a = \eta, \alpha \geq \beta_\alpha = \xi.$$

Then A, R, and S are all in $V^{(B_\alpha)}$ and reasoning similar to that used to prove (3) shows that

$$V^{(B_\alpha)} \models \langle A, R \rangle \ \textit{is a Boolean algebra satisfying } \text{ccc}.$$

We claim that an S-complete ultrafilter in $\langle A, R \rangle$ was added in the passage from $V^{(B_\alpha)}$ to $V^{(B_{\alpha+1})}$.

Now $B_{\alpha+1} = B_\alpha \otimes C_\alpha$, so by Theorem 6.20 the object $U^+ \in V^{(B_{\alpha+1})}$ satisfies

$$V^{(B_{\alpha+1})} \models U^+ \ \textit{is a } P^{(B_\alpha)}(C_\alpha)\textit{-complete ultrafilter in } C_\alpha.$$

But, by construction we have

$$V^{(B_\alpha)} \models C_\alpha \ \textit{is the completion of } \langle A, R \rangle$$

and since $V^{(B_{\alpha+1})} \models S \subseteq P^{(B_\alpha)}(C_\alpha)$ it follows that

$$V^{(B_{\alpha+1})} \models U^+ \cap A \text{ is an } S\text{-complete ultrafilter in } \langle A, R \rangle,$$

as required.

Since $B_{\alpha+1}$ is a complete subalgebra of B, we get

$$V^{(B_\alpha)} \models U^+ \cap A \text{ is an } S\text{-complete ultrafilter in } \langle A, R \rangle,$$

and the proof is complete. $\square$

Corollary 6.34 *If* ZF *is consistent, so is* ZFC $+$ SH$(+\neg$CH$)$.

Proof Assuming GCH, take $\kappa = \aleph_2$ in Theorem 6.33. We get B such that $V^{(B)} \models$ MA $+ 2^{\aleph_0} = \aleph_2$, whence, by Theorem 6.9, $V^{(B)} \models$ SH. The result now follows from Theorem 1.19. $\square$

We conclude this chapter with some remarks on further results about SH that have been obtained. Jensen has shown that $V = L \rightarrow \neg$ SH, which of course yields another proof of the independence of SH. For an account of this proof, see Devlin (1977). Jensen has also shown that SH $+$ GCH is relatively consistent with ZFC, but the proof is very involved: see Devlin and Johnsbraten (1974).

For more applications of Martin's axiom, see Rudin (1977).

Problems

6.35 (The iteration theorem) Let M be a transitive $\in$-model of ZFC, let B be a complete Boolean algebra in the sense of M, let $C, \leq_C$ be elements of $M^{(B)}$ such that $M^{(B)} \models \langle C, \leq_C \rangle$ *is a complete Boolean algebra*, and let $D = B \otimes C$. We identify B as a complete subalgebra of D.

(i) Let F be an M-generic ultrafilter in B and the let i_F be the canonical map of $M^{(B)}$ onto $M[F]$. Then $i_F(c)$ is a complete Boolean algebra (with partial ordering $i_F(\leq_C)$) in $M[F]$. Show that $i_F|D$ is an M-complete homomorphism of D onto $i_F(C)$, that is, preserves the join of any subset of D, which is at the same time a member of M. (Clearly $i_F[D] \subseteq i_F(C)$. To prove equality, use Lemma 1.32. To show that i_F is M-complete, use the M-genericity of F in the form: $A \subseteq F, A \in M \rightarrow \bigwedge A \in F$.)

(ii) *A double generic extension is equivalent to a single one.* Let F be an M-generic ultrafilter in B, and let G be an $M[F]$-generic ultrafilter in $i_F(C)$. Put $H = i_F^{-1}[G] \cap D$. Show that H is an M-generic ultrafilter in D and that $M[H] = M[F][G]$. (Use (i).)

(iii) Let H be an M-generic ultrafilter in D. Then $F = B \cap H$ is an M-generic ultrafilter in B. Put $G = i_F[H]$; show that G is an $M[F]$-generic ultrafilter in $i_F(C)$ and that $M[H] = M[F][G]$. (Use Theorem 6.20 and the remarks following it.)

6.36 (More on $\otimes$) Let B be an complete Boolean algebra and suppose that $V^{(B)} \models \langle C, \leq_C \rangle$ *is a Boolean algebra.* Put $A = B \otimes C$ and identify B as a subalgebra of A (see Lemma 6.15). Then $V^{(B)} \models \hat{A}$ *and* $\hat{B}$ *are Boolean algebras and* $\hat{B}$ *is a subalgebra of* $\hat{A}$.

  (i) Show that, for $a \in A$, we have $[\![a = 1_C]\!] = \bigvee\{b \in B : b \leq a\}$. (Use (6.19.)

  (ii) Define $h \in V^{(B)}$ by $h = \{\langle \hat{a}, a\rangle^{(B)} : a \in A\} \times \{1\}$. Show that $V^{(B)} \models h$ *is a homomorphism of* $\hat{A}$ *onto* C. (Note that $[\![h(\hat{a}) = a]\!] = 1$ for $a \in A$.)

  (iii) Let U_0 be the canonical generic ultrafilter in $\hat{B}$ and, in $V^{(B)}$, let $F = \{x \in \hat{A} : \exists y \in U_0[y \leq x]\}^{(B)}$ be the filter in $\hat{A}$ generated by U_0. Show that $V^{(B)} \models \forall x \in A[h(x) = 1 \leftrightarrow x \in F]$ and deduce that $V^{(B)} \models C \cong \hat{A}/F$. (Use (i).)

  (iv) Let $U^+ \in V^{(A)}$ be the ultrafilter in C defined before (6.18) (*cf.* Theorem 6.20) and $U_* \in V^{(A)}$ be the canonical generic ultrafilter in $\hat{A}$. Show that $V^{(A)} \models h^{-1}[U^+] = U_*$ and deduce that $V^{(A)} \models U_* \in V^{(B)}[U^+]$.

6.37 (The operation inverse to $\otimes$) Let A be a complete Boolean algebra and let B be a complete subalgebra of A. Working inside $V^{(B)}$, let F be the filter in $\hat{A}$ generated by the canonical generic ultrafilter in $\hat{B}$ (*cf.* Problem 6.35(iii)), let $A * B = \hat{A}/F$ and let $\pi : \hat{A} \to A * B$ be the natural epimorphism. Finally, let $p : A \to V^{(B)}$ be defined by $[\![p(a) = \pi(\hat{a})]\!] = 1$ for $a \in A$.

  (i) Show that, for $x, y \in A, b \in B, b \wedge x \leq b \wedge y \leftrightarrow b \leq [\![p(x) \leq p(y)]\!]$ and $b \wedge x = b \wedge y \leftrightarrow b \leq [\![p(x) = p(y)]\!]$. (Like Lemma 6.16.)

  (ii) Let $t \in V^{(B)}$ satisfy $[\![t \in A * B]\!] = 1$. Show that $[\![t = p(a)]\!] = 1$ for some $a \in A$. (Take a to be a suitable mixture of members of A, and use (i).)

  (iii) Show that $[\![A * B \text{ is complete}]\!] = 1$. (Given $X \in V^{(B)}$ such that $[\![X \subseteq A * B \wedge 0 \in X]\!] = 1$, let $a = \bigvee\{x \in A : [\![p(x) \in X]\!] = 1\}$; use (i) and (ii) to show that $[\![a = \bigvee X]\!] = 1$.)

  (iv) Show that $A \cong B \otimes (A * B)$. (Show that p does the trick.)

  (v) Show, inversely, that if

$$V^{(B)} \models C \text{ is a complete Boolean algebra,}$$

then $C \cong (B \otimes C) * B$. (Use Problem 6.36(iii).)

6.38 (Injective Boolean algebras) A Boolean algebra B is said to be *injective* if, for any Boolean algebra A, any homomorphism of a subalgebra of A into B can be extended to the whole of A. B is said to be an *absolute subretract* if whenever

B is a subalgebra of a Boolean algebra A there is an epimorphism from A to B which is the identity on B.

(i) Let B be a complete Boolean algebra and let U_* be the canonical generic ultrafilter in $V^{(B)}$. Show that the following conditions are equivalent:

(a) B is injective;

(b) B is an absolute subretract;

(c) for any Boolean algebra A of which B is a subalgebra, there is $U \in V^{(B)}$ such that $V^{(B)} \models U$ *is an ultrafilter in (the Boolean algebra)* $\hat{A}$ *and* $U_* \subseteq U$;

(d) for any $C \in V^{(B)}$ such that $V^{(B)} \models C$ *is a Boolean algebra*, there is $U \in V^{(B)}$ such that $V^{(B)} \models U$ *is an ultrafilter in* C. (For (iii) $\rightarrow$ (iv), use Problem 6.36(iii).)

(ii) Deduce the *Sikorski Extension Theorem*: any complete Boolean algebra is injective.

7

BOOLEAN-VALUED ANALYSIS

In this chapter we give a brief introduction to real analysis in two types of Boolean-valued model: those arising from *measure algebras,* and those arising from *algebras of projections on Hilbert space.* For both of these models, facts about real numbers holding in the model can be 'externalized' so as to yield results about the mathematical objects to which these real numbers correspond in V: in the first of these models, to measurable functions, and in the second, to self-adjoint operators. We shall assume some familiarity with measure theory and the theory of Hilbert spaces.

We begin with some general considerations on real numbers in Boolean-valued models.

The ordered field of real numbers can be characterized as the unique complete ordered field, up to isomorphism. In set theory there are a number of ways of constructing such a field, for example, by Cauchy sequences or Dedekind cuts in the ordered field $\mathbb{Q}$ of rationals. The results are, of course, all order-isomorphic and the symbol $\mathbb{R}$ used to denote any one of them.

Now let B be a complete Boolean algebra and let $\mathbb{R}^{(B)} \in V^{(B)}$ satisfy

$$V^{(B)} \models \mathbb{R}^{(B)} \textit{ is a complete ordered field.}$$

Let $\mathbb{R}_B$ be a core for $\mathbb{R}^{(B)}$. We may assume that $\hat{r} \in \mathbb{R}_B$ for any $r \in \mathbb{R}$, and in particular, writing $\mathbf{0}$ for the zero element of $\mathbb{R}$, that $\hat{\mathbf{0}}$ is the zero of $\mathbb{R}_B$.

$\mathbb{R}_B$ carries the structure of an *ordered ring.* The operations in this ring are defined in the obvious way: for $u, v \in \mathbb{R}_B$, $u + v$ (resp. uv) is the unique $w \in \mathbb{R}_B$ such that $V^{(B)} \models w = u + v$ (resp. $V^{(B)} \models w = uv$). It is also an *algebra* over $\mathbb{R}$ in which, for $r \in \mathbb{R}$, $r \cdot u$ is the unique $w \in \mathbb{R}_B$ such that $V^{(B)} \models w = \hat{r}u$. Order in $\mathbb{R}_B$ is defined by $u \leq v$ iff $V^{(B)} \models u \leq v$.

Note that, while $\mathbb{R}_B$ is not a field, it is 'almost' a field in the sense that an element $x \in \mathbb{R}_B$ is invertible iff $[\![x = \hat{\mathbf{0}}]\!] = 0_B$.

Boolean-valued models built from measure algebras

A *σ-algebra* is a Boolean algebra in which every countable subset has a join and a meet. A *σ-ideal* in a σ-algebra is an ideal containing the joins of each of its countable subsets. It is readily shown that the quotient of a σ-algebra by a σ-ideal is itself a σ-algebra.

A σ-*field* of subsets of a set E is a field of subsets of E which is closed under countable unions. By a *measure space* we mean a triple $(E, \mathscr{S}, \mu)$ in which $\mathscr{S}$ is a σ-field of subsets of a set E and μ is a σ-finite measure on $\mathscr{S}$, that is, a function on $\mathscr{S}$ taking values in $\mathbb{R}^+ \cup \{\infty\}$ such that (i) $\mu(\varnothing) = 0$; (ii) $\mu\left(\bigcup_{n=1}^{\infty} X_n\right) = \sum_{n=1}^{\infty} \mu(X_n)$ for any countable *disjoint* subfamily $\{X_n : n \in \omega\}$ of $\mathscr{S}$, (iii) there exists a countable disjoint subfamily $\{U_n : n \in \omega\}$ of $\mathscr{S}$ such that for every n, $\mu(U_n) < \infty$ and $\bigcup_{n=1}^{\infty} U_n = E$.

Let $(E, \mathscr{S}, \mu)$ be a measure space. The family $\mathscr{I} = \{X \in \mathscr{S}. \mu(X) = 0\}$ is readily seen to be a σ-ideal in $\mathscr{S}$, so that the quotient $\mathscr{B} = \mathscr{S}/\mathscr{I}$ is a σ-algebra. The ideal $\mathscr{I}$ is the *ideal of sets of μ-measure 0* and $\mathscr{B}$ the *reduced measure algebra* of E. In $\mathscr{B}$, the operations of meet, join and complement are represented by their set-theoretic counterparts. Also equality of two elements $b_1 = S_1/\mathscr{I}$, $b_2 = S_2/\mathscr{I}$ of $\mathscr{B}$ means that S_1 and S_2 differ only by a *set of measure zero, that is, they are almost equal.* It can be shown (see, for example, Halmos 1965, section 7, exercise 4) that $\mathscr{B}$ satisfies the ccc. Since a σ-algebra satisfying the ccc is necessarily complete (see Halmos 1963, section 14, corollary to lemma 1), $\mathscr{B}$ is complete.

We shall need to describe *partitions of unity* in $\mathscr{B}$ (see Ch. 1). If $\{S_i : i \in \omega\}$ is a countable partition of unity in $\mathscr{S}$, then, writing $b_i = S_i/\mathscr{I}$, $\{b_i : i \in \omega\}$ is a partition of unity in $\mathscr{B}$. Using the fact that $\mathscr{B}$ satisfies the ccc, it is not hard to show that every partition of unity in $\mathscr{B}$ arises in this way.

We now introduce the Boolean extension $V^{(\mathscr{B})}$. The natural numbers in $V^{(\mathscr{B})}$ are then mixtures of the form $\sum_{k \in \omega} b_k \cdot \hat{n}_k$ where $\{n_k : k \in \omega\}$ is a set of natural numbers and $\{b_k : k \in \omega\}$ is a partition of unity in $\mathscr{B}$. Similarly, the rational numbers in $V^{(\mathscr{B})}$ are of the form $\sum_{k \in \omega} b_k \cdot \hat{q}_k$ where $\{q_k : k \in \omega\}$ is a set of rational numbers and $\{b_k : k \in \omega\}$ is a partition of unity in $\mathscr{B}$. If $b_k = S_k/\mathscr{I}$ as above, then $\sum_{k \in \omega} b_k \cdot \hat{n}_k$ or $\sum_{k \in \omega} b_k \cdot \hat{q}_k$ may be identified with a step function taking the value n_k or q_k on each S_k.

In order to identify the field $\mathbb{R}^{(\mathscr{B})}$ of *real numbers in $V^{(\mathscr{B})}$* we need to specify a particular construction of $\mathbb{R}$ in V. For our present purposes we shall take $\mathbb{R}$ to consist of the *lower* sections (minus their end points) of Dedekind cuts in $\mathbb{Q}$. Thus we shall define 'r is a real number' as

$$r \subseteq \mathbb{Q} \wedge \exists x \in \mathbb{Q}(x \in r) \wedge \exists x \in \mathbb{Q}(x \notin r) \wedge \forall x \in \mathbb{Q}[x \in r \leftrightarrow \exists y \in \mathbb{Q}(x < y \wedge y \in r)].$$

If we write the above formula as $\varphi(r)$, then $\mathbb{R}^{(\mathscr{B})}$ may be identified as the $\mathscr{B}$-set $\{r \in P\mathbb{Q} : \varphi(r)\}^{(\mathscr{B})}$, and $\mathbb{R}_{\mathscr{B}}$ as a core for this set.

With operations defined in the obvious way, the set of measurable real-vaued functions on E also forms an algebra $\mathbf{M}(E)$. The subset I of functions that are zero almost everywhere, that is, functions f for which $\{x : f(x) \neq 0\}$ has measure 0, is an ideal in $\mathbf{M}(E)$. The following theorem now states the precise sense in which the real numbers in $V^{(\mathscr{B})}$ may be identified with measurable functions on E.

Theorem 7.1 *The algebra* $\mathbb{R}_{\mathscr{B}}$ *is isomorphic to the quotient algebra* $\mathbf{M}(E)/I$.

Sketch of proof Given $u \in \mathbb{R}_{\mathscr{B}}$, for each $q \in \mathbb{Q}$ let

$$(*) \qquad\qquad b_q = [\![\hat{q} \in u]\!].$$

Then from the above definition of real number it follows easily that the three conditions below hold (with meets and joins calculated in $\mathscr{B}$):

$$(1) \quad \bigwedge_{q \in \mathbb{Q}} b_q = 0_{(\mathscr{B})}, \quad (2) \quad \bigvee_{q \in \mathbb{Q}} b_q = 1_{(\mathscr{B})}, \quad (3) \quad b_q = \bigvee_{q < q'} b_{q'}.$$

And conversely $\{b_q \colon q \in \mathbb{Q}\} \subseteq \mathscr{B}$ satisfying (1), (2), and (3) determines an element $u \in \mathbb{R}_{\mathscr{B}}$ such that $\forall q \in \mathbb{Q} \quad b_q = [\![\hat{q} \in u]\!]$.

Now define the map $h \colon \mathbf{M}(E) \to \mathbb{R}_{\mathscr{B}}$ as follows. Given $f \in \mathbf{M}(E)$, for $q \in \mathbb{Q}$ let

$$b_q = \{x \in E \colon q < f(x)\}/\mathscr{I}.$$

It is readily checked that $\{b_q \colon q \in \mathbb{Q}\}$ satisfies (1)–(3) and so determines an element u_f of $\mathbb{R}_{\mathscr{B}}$. We set $h(f) = u_f$.

It is then not difficult to show that h is an algebra homomorphism with kernel I. That h is onto $\mathbb{R}_{\mathscr{B}}$ may be seen by starting with $u \in \mathbb{R}_{\mathscr{B}}$ and considering the associated b_q. If $b_q = S_q/\mathscr{I}$, then since $\{b_q \colon q \in \mathbb{Q}\}$ satisfies (1)–(3), we may take the S_q to satisfy the conditions:

$$(1') \quad \bigcap_{q \in \mathbb{Q}} S_q = \varnothing, \quad (2') \quad \bigcup_{q \in \mathbb{Q}} S_q = E, \quad (3') \quad S_q = \bigcup_{q < q'} S_{q'}.$$

Now define $f_u \colon E \to \mathbb{R}$ by

$$f_u(x) = \sup\{q \colon x \in S_q\}.$$

One checks that $f_u \in M(E)$ and that

$$S_q = \{x \in E \colon q < f_u(x)\}.$$

From this it follows easily that $h(f_u) = u$, establishing the surjectivity of h.

So h is an homomorphism of $\mathbf{M}(E)$ onto $\mathbb{R}_{\mathscr{B}}$ with kernel I. The theorem follows. $\qquad\qquad\square$

A property $\varphi(x)$ defined on E is said to hold *almost everywhere* if $\{x \in E \colon \neg\varphi(x)\}$ has measure 0. Roughly speaking, truth in $V^{(\mathscr{B})}$ corresponds to truth in V almost everywhere. A simple illustration of this is the following. Let us say that $f \in \mathbf{M}(E)$ is a *correlate* of $u \in \mathbb{R}_{\mathscr{B}}$ if $h(f) = u$. Two measurable functions f and g are then correlates of a single member in $\mathbb{R}_{\mathscr{B}}$ iff $f = g$ almost everywhere, that is, if $\{x \in E: f(x) \neq g(x)\}$ has measure 0. Similarly, f and g are correlates respectively of u, v for which $u < v$ precisely when $f < g$ holds almost everywhere. The fact that *the least upper bound principle* for the real numbers holds in $V^{(\mathscr{B})}$ then immediately yields the following.

Theorem 7.2 *Let U be a nonempty set of measurable functions and suppose that a measurable function f exists such that $g \leq f$ almost everywhere for all $g \in U$. Then there exists a measurable function h such that:*

1. *$g \leq h$ almost everywhere for all $g \in U$;*
2. *if k is a measurable function such that $g \leq k$ almost everywhere for all $g \in U$, then $h \leq k$ almost everywhere.*

Before stating our final result we require

Lemma 7.3 *Suppose that $V^{(\mathscr{B})} \models u: \omega \to \mathbb{R}^{(\mathscr{B})}$ and, for each $k \in \omega$, $u_k \in \mathbb{R}_{\mathscr{B}}$ satisfies $[\![u_k = u(\hat{k})]\!] = 1$. For each $k \in \omega$, let f_k be a correlate of u_k and let g be a correlate of $v \in \mathbb{R}_B$. Then the following are equivalent:*

- *$(f_k)_{k \in \omega}$ converges to g almost everywhere*
- *$V^{(\mathscr{B})} \models \lim_{k \to \infty} u(k) = v$.*

Proof We have

$$[\![\lim_{k \to \infty} u(k) = v]\!] = 1 \leftrightarrow [\![\forall \varepsilon > 0 \exists n \forall k \geq n |v - u(k)| < \varepsilon]\!] = 1$$

$$\leftrightarrow \bigcap_{\varepsilon > 0} \bigcup_{n} \bigcap_{k \geq n} [\![|v - u(k)| < \hat{\varepsilon}]\!] = 1$$

$$\leftrightarrow \bigcap_{\varepsilon > 0} \bigcup_{n} \bigcap_{k \geq n} [\![|v - u_k| < \hat{\varepsilon}]\!] = 1$$

$$\leftrightarrow (\bigcap_{\varepsilon > 0} \bigcup_{n} \bigcap_{k \geq n} \{x : |g(x) - f_k(x)| < \varepsilon\})/\mathscr{I} = E/\mathscr{I}$$

$$\leftrightarrow \{x : \forall \varepsilon > 0 \exists n \forall k \geq n| \, |g(n) - f_k(n)| < \varepsilon\}/\mathscr{I} = E/\mathscr{I}$$

$$\leftrightarrow (f_k)_{k \in \omega} \text{ converges to } g \text{ almost everywhere.} \qquad \square$$

Now suppose that $V^{(\mathscr{B})} \models u: \omega \to \mathbb{R}^{(\mathscr{B})}$ and, for each $k \in \omega$, $u_k \in \mathbb{R}_{\mathscr{B}}$ satisfies $[\![u_k = u(\hat{k})]\!] = 1$. For each $k \in \omega$, let f_k be a correlate of u_k. Let $m = \sum_k b_k \cdot \widehat{n_k}$ be a natural number in $V^{(\mathscr{B})}$, that is, $n_k \in \omega$ and $\{b_k: k \in \omega\}$ is a partition of unity in $\mathscr{B}$. Let v be the unique member of $\mathbb{R}_{\mathscr{B}}$ for which $[\![v = u(m)]\!] = 1$. Let $\{S_k: k \in \omega\} \subseteq \mathscr{S}$ satisfy $b_k = S_k/\mathscr{I}$; we may assume that $\{S_k: k \in \omega\}$ is a partition of E. Finally let g be the function on E whose restriction to each S_k is f_{n_k}; it is not difficult to check that g is a correlate of v. From this,

Lemma 7.3, and the fact that the Bolzano–Weierstrass theorem holds for $\mathbb{R}_{\mathscr{B}}$ with probability 1, one easily deduces the following.

Theorem 7.4 *Suppose that g, f_0, f_1, ... are measurable functions with $|f_k| < g$ almost everywhere for all k. Then there exists a measurable function h such that $|h| < g$ almost everywhere and h is a Boolean-valued cluster function for $(f_k)_{k\in\omega}$, that is, for every $\varepsilon > 0$ and every $m \in \omega$ there exist a measurable function e, a sequence $(n_k)_{k\in\omega}$ of natural numbers, and a partition $\{S_k : k \in \omega\}$ of E such that*

1. *$m \leq n_k$ for every k;*
2. *for each k, the restriction of e to S_k is f_{n_k};*
3. *$|e(x) - h(x)| < \varepsilon$ almost everywhere.*

Boolean-valued models built from algebras of projections

Let $\mathscr{H}$ be a (complex) Hilbert space, which will remain fixed throughout our discussion: all operators, etc. will be assumed to be defined on $\mathscr{H}$. Addition, subtraction, and composition of operators will be denoted by $+$, $-$, and juxtaposition. The identity operator will be denoted by I and the zero operator by 0: these are defined by $Ix = x$ and $0x = 0$. We recall that a *projection* is a bounded self-adjoint operator P such that $P^2 = P$. The range $\mathrm{ran}(P) = \{x : Px = x\}$ of a projection is a closed linear subspace of $\mathscr{H}$; conversely each closed linear subspace S determines a projection P_S such that $S = \mathrm{ran}(P_S)$. The set $\mathscr{P}$ of projections may be partially ordered by defining

$$P \leq Q \leftrightarrow \mathrm{ran}(P) \subseteq \mathrm{ran}(Q) \leftrightarrow PQ = QP = P.$$

With this partial ordering $\mathscr{P}$ is a complete lattice in which I and 0 are the top and bottom elements. Joins and meets in $\mathscr{P}$ are given by:

$$\bigvee_{i\in I} P_i = P_S \text{ with } S = \text{closed linear span of } \bigcup_{i\in I} ran(P_i)$$

$$\bigwedge_{i\in I} P_i = P_S \text{ with } S = \bigcap_{i\in I} ran(P_i)$$

Each element $P \in \mathscr{P}$ has a complement P^c in $\mathscr{P}$ given by $P^c = I - P$.[1] Clearly $P^{cc} = P$ and $P \leq Q$ iff $Q^c \leq P^c$. While the lattice $\mathscr{P}$ is not distributive, it is *orthomodular*, that is, satisfies the following condition:

$$P \leq Q \rightarrow P = Q \wedge (P \vee Q^c).$$

---

[1] We denote complementation in $\mathscr{P}$ by ^{c} in order to avoid confusion with the customary use in functional analysis of the asterisk * to denote the adjoint operator.

The following laws governing complements of joins and meets in $\mathscr{P}$ obtain:

$$\left(\bigvee_{i \in I} P_i\right)^c = \left(\bigwedge_{i \in I} P_i^c\right), \quad \left(\bigwedge_{i \in I} P_i\right)^c = \left(\bigvee_{i \in I} P_i^c\right).$$

In sum, $\mathscr{P}$ is what is known as a *complete orthomodular ortholattice*.

Projections $P, Q \in \mathscr{P}$ are said to *commute* if $PQ = QP$. The join and meet of any commuting pair $P, Q \in \mathscr{P}$ are given by

$$P \vee Q = P + Q - PQ, \quad P \wedge Q = PQ.$$

It can be shown (Jauch 1968, section 5–7) that $P, Q \in \mathscr{P}$ commute if and only if they are *compatible*, that is,

$$P = (P \wedge Q) \vee (P \wedge Q^c) \text{ and } Q = (Q \wedge P) \vee (Q \wedge P^c).$$

Clearly, P is compatible with Q iff P^c is compatible with Q. It is easily seen that the compatibility of P and Q is equivalent to the assertion that the sublattice of $\mathscr{P}$ generated by $\{P, Q, P^c, Q^c\}$ is a Boolean algebra. It follows that, for any subset $\mathscr{X}$ of $\mathscr{P}$ consisting of pairwise compatible elements, the sublattice of $\mathscr{P}$ generated by $\mathscr{X} \cup \{P^c: P \in \mathscr{X}\}$ is a Boolean algebra. It can also be shown (Jauch 1968, section 5–8), that, if P is compatible with each Q_i for $i \in I$, then P is compatible with both $\bigvee_{i \in I} Q_i$ and $\bigwedge_{i \in I} Q_i$.

A *complete Boolean projection algebra* is a subset $\mathscr{B}$ of $\mathscr{P}$ such that

1. The elements of $\mathscr{B}$ are pairwise compatible, hence commute;
2. Both I and 0 are elements of $\mathscr{B}$; if $P \in \mathscr{B}$, then $P^c \in \mathscr{B}$, and if $\{P_i: i \in I\} \subseteq \mathscr{B}$, then both $\bigvee_{i \in I} P_i$ and $\bigwedge_{i \in I} P_i \in \mathscr{B}$.

Clearly, if $\mathscr{B}$ is a complete Boolean projection algebra, then $(\mathscr{B}, \leq)$ is a complete Boolean algebra with top and bottom elements I and 0, respectively.

Lemma 7.5 *Any set of pairwise commuting projections is contained in a complete Boolean projection algebra.*

Proof Let $\mathscr{X}$ be a set of pairwise commuting projections. Using Zorn's lemma, there is a maximal set $\mathscr{M}$ of pairwise commuting projections such that $\mathscr{X} \subseteq \mathscr{M}$. Then $\mathscr{M}$ is a complete Boolean projection algebra. For I and 0 are both compatible with every member of $\mathscr{M}$, hence members of $\mathscr{M}$ by maximality. If $P \in \mathscr{M}$, then P^c is compatible with every member of $\mathscr{M}$, hence also in $\mathscr{M}$ by maximality. Finally, for any subset $\{P_i: i \in I\}$ of $\mathscr{M}$, both $\bigvee_{i \in I} P_i$ and $\bigwedge_{i \in I} P_i$ are compatible with every member of $\mathscr{M}$; maximality again implies that they are both members of $\mathscr{M}$. $\qquad\square$

We shall need a few facts about *spectral resolutions* of self-adjoint operators. A *spectral family* is a subset $\{E_\lambda: \lambda \in \mathbb{R}\}$ of $\mathscr{P}$ satisfying the following conditions:

$$(1) \quad \bigwedge_\lambda E_\lambda = 0 \quad (2) \quad \bigvee_\lambda E_\lambda = I \quad (3) \quad \text{For any } \lambda \in \mathbb{R}, \ E_\lambda = \bigwedge_{\lambda < \mu} E_\mu.$$

It follows from (3) that the members of a spectral family are pairwise commuting. The *spectral theorem* (see Jauch 1968, section 4–3) asserts that corresponding to each self-adjoint operator A there is a unique spectral family $\{E_\lambda : \lambda \in \mathbb{R}\}$ for which

$$A = \int \lambda dE_\lambda.$$

The spectral family $\{E_\lambda : \lambda \in \mathbb{R}\}$ is called the *spectral resolution* of A.

Let A and A' be self-adjoint operators with spectral resolutions $\{E_\lambda : \lambda \in \mathbb{R}\}$, $\{E'_\lambda : \lambda \in \mathbb{R}\}$. A and A' are said to be *commutable* if $E_\lambda E'_\mu = E'_\mu E_\lambda$ for all λ, μ. When A and A' are bounded (continuous), commutability is equivalent to commutativity: $AA' = A'A$.

Now let $\mathscr{B}$ be a complete Boolean projection algebra. We define the *closure* of $\mathscr{B}$ to be the set $\overline{\mathscr{B}}$ comprising all those self-adjoint operators whose spectral resolutions are included in $\mathscr{B}$. Lemma 7.5 now immediately yields the following.

For any set $\mathscr{X}$ of self-adjoint pairwise commutable operators there exists a complete Boolean projection algebra whose closure includes $\mathscr{X}$.

We now fix a complete Boolean projection algebra $\mathscr{B}$ and turn our attention to the $\mathscr{B}$-valued universe $V^{(\mathscr{B})}$. The natural numbers and rational numbers in $V^{(\mathscr{B})}$ are, respectively, mixtures of the form $\sum_{k \in \omega} P_k \cdot \hat{n}_k$ or $\sum_{k \in \omega} P_k \cdot \hat{q}_k$ with $\{P_k : k \in \omega\}$ a partition of unity in $\mathscr{B}$ and $\{n_k : k \in \omega\}$, $\{q_k : k \in \omega\}$ sets of natural and rational numbers, respectively. Straightforward rules for addition, multiplication, and order of these mixtures apply. For example, if $u = \sum_{k \in \omega} P_k \cdot \hat{n}_k$ and $v = \sum_{k \in \omega} Q_k \cdot \hat{m}_k$, then, noting that $\{P_k Q_j : k, j \in \omega\}$ is a partition of unity in $\mathscr{B}$, we have

$$u + v = \sum_{k,j \in \omega} P_i Q_j \cdot \widehat{n_k + m_j}, \quad uv = \sum_{k,j \in \omega} P_i Q_j \cdot \widehat{n_k m_j}, \quad [\![u < v]\!] = \bigvee \{P_k Q_j : n_k < m_j\}.$$

In order to identify the field $\mathbb{R}^{(\mathscr{B})}$ of *real numbers in* $V^{(\mathscr{B})}$ we again need to specify a particular construction of $\mathbb{R}$ in V. In this situation we shall take $\mathbb{R}$ to consist of the *upper* sections (now including end points, if any) of Dedekind cuts in $\mathbb{Q}$. Thus we shall define 'r is a real number' as

$$r \subseteq \mathbb{Q} \wedge \exists x \in \mathbb{Q}(x \in r) \wedge \exists x \in \mathbb{Q}(x \notin r) \wedge \forall x \in \mathbb{Q}[x \in r \leftrightarrow \forall y \in \mathbb{Q}(x < y \rightarrow y \in r)].$$

If we write the above formula as $\varphi(r)$, then $\mathbb{R}^{(\mathscr{B})}$ may be identified as the $\mathscr{B}$-set $\{r \in P\mathbb{Q} : \varphi(r)\}^{(\mathscr{B})}$, and $\mathbb{R}_{\mathscr{B}}$ as a core for this set.

Given $u \in \mathbb{R}_{\mathscr{B}}$, for $q \in \mathbb{Q}$ define $P_q \in \mathscr{B}$ to be $[\![\hat{q} \in u]\!]$. It is then straightforward to verify that $\bigwedge_{q \in \mathbb{Q}} P_q = 0$, $\bigvee_{q \in \mathbb{Q}} P_q = I$, and for all $q \in \mathbb{Q}$, $P_q = \bigwedge_{q < s} P_s$. From this it follows easily that if for each $\lambda \in \mathbb{R}$ we define $E_\lambda = \bigwedge_{\lambda < q} P_q$, then $\{E_\lambda : \lambda \in \mathbb{R}\}$ is a spectral family $\subseteq \mathscr{B}$.

Reciprocally, given a spectral family $\{E_\lambda : \lambda \in \mathbb{R}\} \subseteq \mathscr{B}$, define $u \in V^{(\mathscr{B})}$ by $\mathrm{dom}(u) = \{\hat{q} : q \in \mathbb{Q}\}$ and $u(\hat{q}) = P_q$. It is then easy to check that $[\![u \in \mathbb{R}^{(\mathscr{B})}]\!] = 1$.

The correspondence thus established between $\mathbb{R}_{\mathscr{B}}$ and spectral families $\subseteq \mathscr{B}$ is bijective. Since there is also a bijective correspondence between $\overline{\mathscr{B}}$ and the collection of spectral families $\subseteq \mathscr{B}$, it follows that there is a bijective correspondence between $\mathbb{R}_{\mathscr{B}}$ and $\overline{\mathscr{B}}$. This may be informally expressed by saying that the 'interpretation' of a real number in $V^{(\mathscr{B})}$ is a self-adjoint operator in $\overline{\mathscr{B}}$.

In Takeuti (1978) it is shown that the bijective correspondence between $\mathbb{R}_{(\mathscr{B})}$ and $\overline{\mathscr{B}}$ is actually an isomorphism of $\mathbb{R}$-algebras. There one will also find results obtained by 'interpreting' in $V^{(\mathscr{B})}$ a number of theorems of elementary analysis. For instance, in the case of the intermediate value theorem one obtains:

Suppose that $f \colon \mathbb{R} \to \mathbb{R}$ is continuous. Let A and B be commutable self-adjoint operators with $A \le B$, Y a self-adjoint operator commutable with both A and B for which $f(A) \le Y \le B$. Then there exists a self-adjoint operator X commutable with A, B, and Y such that $A \le X \le B$ and $Y = f(X)$.

Davis (1977) employs the Boolean-valued analysis just outlined to provide a novel interpretation of the formalism of quantum theory. He points out that in 'quantizing' a classical theory the symbols representing real quantities are replaced by symbols representing corresponding self-adjoint operators on Hilbert space. The correspondence, for a complete Boolean projection algebra $\mathscr{B}$, between self-adjoint operators in $\overline{\mathscr{B}}$ and $\mathscr{B}$-valued real numbers suggests that such projection algebras (or the associated Boolean-valued universes) be regarded as *reference frames* with respect to which measurements may be made of the observables corresponding to the operators in $\overline{\mathscr{B}}$. Under this interpretation the $\mathscr{B}$-valued real number correlated with a given observable A is the value that would be obtained by measuring A with respect to the frame $\mathscr{B}$. The key point is that measurements *must* be made relative to a Boolean frame, and, like inertial frames in relativity theory, there is no absolute such frame.

In Davis's interpretation, it is supposed that certain sentences of the language of set theory are given, which express relationships among real physical quantities. Some of these sentences will represent basic physical postulates, others will express the result of a measurement. The correctness of such a sentence σ as a description of reality is embodied in the assertion that $V^{(\mathscr{B})} \models \sigma$ for any complete Boolean projection algebra $\mathscr{B}$. The role played by a particular Boolean frame may be grasped by analyzing a sentence expressing the fact that a real quantity resulting from an interaction with some apparatus satisfies some condition (an inequality, for instance). Such a sentence has the form $p \to q$, where

p expresses the effect of the interaction and q the resulting condition.[2] Then $p \rightarrow q$ will hold, that is, $V^{(\mathscr{B})} \models p \rightarrow q$, in every Boolean frame $\mathscr{B}$. But in order to obtain from this the truth of q, that is, $V^{(\mathscr{B})} \models q$, a frame $\mathscr{B}$ must be chosen in which $V^{(\mathscr{B})} \models p$. An appropriate choice of $\mathscr{B}$ will be determined by the physics of the apparatus.

Quantities that can be measured simultaneously correspond to commuting self-adjoint operators.[3] Hence there is a complete Boolean projection algebra $\mathscr{B}$ such that $\overline{\mathscr{B}}$ contains all such operators: such an algebra $\mathscr{B}$ may be considered a reference frame with respect to which the measurements are being made. In the case of complementary quantities such as position and momentum, the corresponding operators do not commute and so there is no Boolean algebra $\mathscr{B}$ such that $\overline{\mathscr{B}}$ contains them both. A position measurement and a momentum measurement can accordingly only be made with respect to *distinct* frames of reference.

The notorious anomalies of quantum mechanics are thus seen to dissolve under this interpretation. For example, consider the *two-slit experiment*. Here we have two slits in front of a screen equipped with counters, and a beam of particles that can reach the screen only by passing through one of the slits. The pattern on the screen when both slits are open is, anomalously, found not to coincide with the union of the patterns obtained when each is open separately. The anomaly disappears when one observes that the Boolean frame associated with opening both slits is *different* from that associated with the opening of a single slit: the first frame corresponds to a momentum measurement, the second, to a position measurement. In the case of the *Einstein–Podolsky–Rosen paradox*, two particles A and B interact in such a way that the sum of their momenta after the interaction is 0, so that a measurement of the momentum of A will also yield the momentum of B. Such a measurement will accordingly not only cause a 'collapse' of the state function of A, but also, and simultaneously, that of B, which by this time may be very far away. Here the paradox is resolved as follows. Let M_A and M_B be the momentum operators corresponding to A and B, respectively, and let p_A and p_B represent their momenta at some time t. Then, choosing an appropriate Boolean frame $\mathscr{B}$ for which $\overline{\mathscr{B}}$ contains both A and B, we know that $V^{(\mathscr{B})} \models p_A + p_B = 0$. From this we *deduce* that $M_A + M_B = 0$: it is the choice of the frame, not the measurement of p_A, that guarantees the result.

---

[2] Davis offers the example here of a particle's presence in a slit—a 'position' measurement—revealed by the action of a counter. In this case p may be taken to be the statement 'the counter clicks' and q the statement 'the particle was in the slit'.

[3] Note that since the transition from real quantities to self-adjoint operators is effected with respect to a given frame, the fact that a particular observable is represented by some specific operator must also be taken relative to a frame.

8

INTUITIONISTIC SET THEORY AND HEYTING-ALGEBRA-VALUED MODELS

In this final chapter intuitionistic set theory is introduced and the idea of a Boolean-valued model of classical set theory extended to that of a Heyting-algebra-valued model of intuitionistic set theory.

Intuitionistic Zermelo set theory

The system IZ of *intuitionistic Zermelo set theory* is formulated in the language $\mathcal{L}$, subject to the axioms and rules of intuitionistic first-order logic. Arguments in IZ will be presented informally; in particular we shall make use of the standard notations of classical set theory introduced previously: $\exists y \in x$, $\forall y \in x$, $\{x : \varphi\}$, $x \cup y$, Px, $\langle x, y \rangle$, $x \subseteq y$, $\emptyset, 0, 1, 2$, etc. The *axioms* of IZ are *Extensionality, Separation, Union, Power set, Infinity* (all as stated in the Prerequisites), together with

$$Pairing \quad \forall x \forall y \exists z \forall w (w \in z \leftrightarrow w = x \lor w = y).$$

For any set A, PA is a complete Heyting algebra with operations $\cup, \cap$ and $\Rightarrow$, where $U \Rightarrow V = \{x : x \in U \to x \in V\}$, and top and bottom elements A and $\emptyset$, respectively.

We write $\{\tau | \varphi\}$ for $\{x : x = \tau \land \varphi\}$, where τ is a closed term; notice that without the law of excluded middle we cannot conclude that $\{\tau | \varphi\} = \emptyset$ or $\{\tau\}$. From Extensionality we infer that $\{\tau | \varphi\} = \{\tau | \psi\}$ iff $(\varphi \leftrightarrow \psi)$; thus, in particular, the elements of $P1$ (recall that $1 = \{0\}$) correspond naturally to *truth values*, that is, propositions identified under equivalence. $P1$ is called the (Heyting) *algebra of truth values* and is denoted by Ω. The top element 1 of Ω is usually written *true* and the bottom element $\emptyset$ as *false*.

We note that, in IZ, Ω plays the role of a *subset classifier*. That is, for each set A, subsets of A are correlated bijectively to maps $A \to \Omega$. To wit, each subset $X \subseteq A$ is correlated with its *characteristic map* $\chi_X : A \to \Omega$ given by $\chi_X(x) = \{0 | x \in X\}$; conversely each map $f : A \to \Omega$ is correlated with the subset $f^{-1}(1)$ of A.

Properties of Ω correspond to *logical properties* of the set theory. Recall, for instance,

LEM (law of excluded middle) $\varphi \lor \neg\varphi$
WLEM (weakened law of excluded middle) $\neg\varphi \lor \neg\neg\varphi$.

LEM and WLEM correspond respectively to the properties

$$\forall \omega \in \Omega \cdot \omega = \textit{true} \vee \omega = \textit{false} \quad \forall \omega \in \Omega \cdot \omega = \textit{false} \vee \omega \neq \textit{false}.$$

Calling a set A *decidable* if $\forall x \in A \; \forall y \in A(x = y \vee x \neq y)$, each of the following is equivalent, in IZ, to LEM:

1. *Every set is decidable.*
2. Ω *is decidable.*
3. $\Omega = 2$.
4. *Membership is decidable:* $\forall x \forall y (x \in y \vee x \notin y)$.
5. $\forall x (0 \in x \vee 0 \notin x)$.
6. $(2, \leq)$ *is well-ordered.*

(To show that 6. implies LEM, observe that the least element of $\{0|\varphi\} \cup \{1\} \subseteq 2$ is either 0 or 1; if it is 0, φ must hold, and if it is 1, φ must fail.)

Using the axiom of infinity, the set $\mathbb{N}$ of natural numbers can be constructed as usual. $\mathbb{N}$ is decidable and satisfies the familiar Peano axioms including induction, but it is well-ordered only if LEM holds. In fact LEM also follows from the *domino principle* for $\mathbb{N}$:

$$\varphi(0) \wedge \exists n \neg \varphi(n) \rightarrow \exists n [\varphi(n) \wedge \neg \varphi(n+1)].^{1}$$

To see this, take any formula ψ and define $\varphi(n)$ to be the formula $n = 0 \vee (n = 1 \wedge \psi)$. Then clearly $\varphi(0) \wedge \exists n \neg \varphi(n)$ holds, so we infer from the domino principle that there is n_0 for which $\varphi(n)$ and $\neg \varphi(n+1)$, that is,

$$(*) \qquad\qquad n_0 = 0 \vee (n_0 = 1 \wedge \psi)$$

and

$$\neg(n_0 + 1 = 1 \wedge \psi)$$

whence

$$\neg(n_0 = 0 \wedge \psi).$$

From this last we infer $n_0 = 0 \rightarrow \neg\psi$, which, together; with $(*)$, gives $\psi \vee \neg\psi$.

The notion of a *function* is defined as usual in IZ; we employ the standard notations for functions. A *choice function* on a set A is a function f with domain A such that $f(a) \in a$ whenever $\exists x \; x \in a$. The *axiom of choice* AC is the assertion that every set has a choice function. Remarkably, AC implies LEM; in fact *it*

---

[1] Here and in the sequel we shall use lower case letters m, n as variables ranging over $\mathbb{N}$.

is provable in IZ that if each doubleton has a choice function, then LEM holds (and conversely).

To prove this, define $U = \{x \in 2 : x = 0 \vee \varphi\}$ and $V = \{x \in 2 : x = 1 \vee \varphi\}$, and suppose given a choice function f on $\{U, V\}$. Writing $a = f(U), b = f(V)$, we then have $a \in U, b \in V$, that is,

$$(a = 0 \vee \varphi) \wedge (b = 1 \vee \varphi).$$

Hence

$$(a = 0 \wedge b = 1) \vee \varphi$$

whence

$$a \neq b \vee \varphi. \tag{$*$}$$

But

$$\varphi \to U = V \to a = b,$$

so that

$$a \neq b \to \neg\varphi.$$

This, together with $(*)$, gives $\varphi \vee \neg\varphi$.

It can also be shown that the assertion *any singleton has a choice function* is equivalent in IZ to the (intuitionistically invalid) 'independence of premises' rule,

$$\frac{\psi \to \exists x(x \in A \wedge \varphi(x))}{\exists x(\psi \to x \in A \wedge \varphi(x))}.$$

In classical set theory one proves the well-known *Schröder–Bernstein theorem*: if each of two sets A and B can be injected into the other, then there is a bijection between A and B. This is usually derived as a consequence of the proposition

SB: *for any set X and any injection $f : X \to X$ there is a bijection h:*
$X \to X$ *such that* $h \subseteq f \cup f^{-1}$, *that is,*
$\forall x \in X[h(x) = f(x) \vee f(h(x)) = x].$

In IZ this assertion implies (and so is equivalent to) LEM. Here is the proof.

Define, for any formula φ,

$$\mathbb{N}^{\varphi} = \mathbb{N} - \{0\} \cup \{0|\varphi\} \quad f = \{(n, n+1) : n \neq 0\} \cup \{(0, 1)|\varphi\}.$$

Then $f\colon \mathbb{N}^\varphi \to \mathbb{N}^\varphi$. Clearly

$$(*) \qquad\qquad 1 \in \mathrm{range}(f) \leftrightarrow 0 \in \mathbb{N}^\varphi \leftrightarrow \varphi.$$

Now suppose given a bijection $h\colon \mathbb{N}^\varphi \to \mathbb{N}^\varphi$ such that

$$\forall x \in \mathbb{N}^\varphi [h(x) = f(x) \vee f(h(x)) = x].$$

If φ holds, then f is just the usual successor function on $\mathbb{N}(= \mathbb{N}^\varphi)$ and so

$$\varphi \wedge h(n) = 0 \to h(n) \neq f(n) \to 1 = f(0) = f(h(n)) = n \to n = 1,$$

whence

$$\varphi \to h(1) = 0.$$

Thus

$$(**) \qquad\qquad h(1) \neq 0 \to \neg\varphi$$

But

$$h(1) = f(1) \vee f(h(1)) = 1.$$

The first disjunct implies $h(1) \neq 0$ and $(**)$ gives $\neg\varphi$. From the second disjunct we infer $1 \in \mathrm{range}(f)$ and $(*)$ yields φ. Thus we have derived $\varphi \vee \neg\varphi$.

Another theorem of classical set theory that, in IZ, implies the law of excluded middle is the *Stone Representation Theorem*, namely, the assertion that every Boolean algebra is isomorphic to a field of sets.[2] To see this, observe first that, in IZ, the Stone Representation Theorem implies the assertion

$$(*) \qquad \textit{In each Boolean algebra the intersection of the family of all}$$
$$\textit{its prime filters is } \{1\}.$$

For if B is a field of subsets of a set S, then, for each $x \in S, F_x = \{X \in B\colon x \in X\}$ is a prime filter in B. If $X \in \bigcap_{x \in S} F_x$, then $x \in X$ for all $x \in X$, whence $X = S$. Therefore $\bigcap_{x \in S} F_x = \{S\}$ and so B has the property asserted in $(*)$. So if the Stone Representation theorem holds $(*)$ obtains universally.

Now we show that $(*)$ implies LEM in IZ. For each Boolean algebra B, let $\mathrm{Prim}(B)$ be the set of prime filters in B. Then $\bigcap \mathrm{Prim}(B) = \{1\}$ and we have

$$(**) \qquad\qquad \mathrm{Prim}(B) = \emptyset \to B \text{ is trivial.}$$

---

[2] Proved in Ch. 0.

For if B is trivial, it has no filters, so that $\mathrm{Prim}(B) = \emptyset$. Conversely, if $\mathrm{Prim}(B) = \emptyset$, then $\{1\} = \bigcap \mathrm{Prim}(B) = \bigcap \emptyset = B$, so that B is trivial.

Now let φ be any formula, and define

$$B_\varphi = \{\omega \in \Omega : \omega = \varphi \ or \ \omega = true\}.$$

This is easily shown to be a Boolean algebra in which $0 = \varphi$, $1 = true$, meets are conjunctions, joins are disjunctions, and the complement of ω is $(\omega \to \varphi)$. Clearly

$$(\text{***}) \qquad\qquad B_\varphi \text{ is trivial } \leftrightarrow \varphi.$$

Putting (**) and (***) together, we see that

$$\varphi \leftrightarrow \mathrm{Prim}(B_\varphi) = \emptyset \leftrightarrow \neg \exists X \ X \in \mathrm{Prim}(B_\varphi).$$

Thus φ is equivalent to a negated formula, and so satisfies the law of double negation. Since φ was arbitrary, it follows that LDN, and hence LEM, holds generally.

In ZFC one can prove the so-called *order extension principle* to the effect that every partial ordering on a set can be extended to a total ordering. We will show that, in IZ, this principle implies the intuitionistaically invalid law $\varphi \to \psi \vee \psi \to \varphi$.

To prove this, we first observe that if $U, V \subseteq 1$, then

$$(\text{*}) \qquad\qquad (U = 1 \to V = 1) \leftrightarrow U \subseteq V.$$

Now suppose that $\leq$ is a total order on Ω extending $\subseteq$. Then $U \leq 1$ for all $U \in \Omega$. Now

$$U \leq V \wedge U = 1 \to 1 \leq V \to V = 1,$$

whence, using (*),

$$U \leq V \to (U = 1 \to V = 1) \to U \subseteq V.$$

We conclude that $\leq$ and $\subseteq$ coincide. Accordingly, if $\subseteq$ could be extended to a total order on Ω, $\subseteq$ would have to be a total order on Ω itself. But this is clearly tantamount to the truth of $\varphi \to \psi \vee \psi \to \varphi$ for arbitrary formulas φ and ψ.

The negation operation $\neg$ on propositions corresponds to the complementation operation on Ω; we use the same symbol $\neg$ to denote the latter. This operation of course satisfies

$$\omega \subseteq \neg\omega' \leftrightarrow \omega \cap \omega' = \ false.$$

Classically, $\neg$ also satisfies the dual law, viz.

$$\neg\omega \subseteq \omega' \leftrightarrow \omega \cup \omega' = \; true.$$

But intuitionistically, this is far from being the case. Indeed, the assumption that there exists *any* operation $-: \Omega \to \Omega$ satisfying

$$-\omega \subseteq \omega' \leftrightarrow \omega \cup \omega' = true$$

implies (and so is equivalent to) LEM. For suppose such an operation existed. Then

$$-true \subseteq false \leftrightarrow false \cup true = true,$$

so that $-true \subseteq false$, whence $-true = false$. Next,

$$0 \in -\omega \wedge 0 \in \omega \to 0 \in -\omega \wedge \omega = true \to 0 \in -true = false.$$

Since $0 \notin false$, it follows that

$$0 \in -\omega \to 0 \notin \omega \to 0 \in \neg\omega,$$

and from this we infer that $-\omega \subseteq \neg\omega$. Since, obviously, $\omega \cup -\omega = true$, it then follows that, for any $\omega, \omega \cup \neg\omega = true$, which is LEM.

Intuitionistic Zermelo–Fraenkel set theory

Intuitionistic Zermelo–Fraenkel set theory IZF is obtained by adding to IZ the axioms of *replacement* and *regularity*[3] (both as stated in Chapter 1).

It is to be expected that the many classically equivalent definitions of *well-ordering* and *ordinal* become distinct within IZF. The definitions we give here work reasonably well.

Definition   A set x is *transitive* if $\forall y \in x \; y \subseteq x$; an *ordinal* is a transitive set of transitive sets. The class of ordinals is denoted by ORD and we use (italic) letters $\alpha, \beta, \gamma, \ldots$ as variables ranging over it. A transitive subset of an ordinal is called a *subordinal*. An ordinal α is *simple* if $\forall \beta \in \alpha \; \gamma \in \alpha$ $(\beta \in \gamma \vee \beta = \gamma \vee \gamma \in \beta)$.

---

[3]The version of the axiom of regularity we have been working with is commonly known as the $\in$-*induction scheme*. Classically, this is equivalent to the asserts that each nonempty set u has a member x, which is $\in$-*minimal*, that is, for which $x \cap u = \emptyset$. It is easy to see that this implies LEM: an $\in$-minimal element of the set $\{0|\varphi\} \cup \{1\}$ is either 0 or 1; if it is 0, φ must hold, and if it is 1, φ must fail; thus if foundation held we would get $\varphi \vee \neg\varphi$.

Thus, for example, the ordinals $1, 2, 3, \ldots$ as well as the first infinite ordinal ω to be defined below, are all simple. Every subordinal (hence every element) of a simple ordinal is simple. But, in contrast with classical set theory, intuitionistically not every ordinal can be simple, because the simplicity of the ordinal $\{0, \{0|\varphi\}\}$ implies $\varphi \vee \neg\varphi$.

We next state the central properties of ORD.

Definition The *successor* α^+ of an ordinal α is $\alpha \cup \{\alpha\}$; the *supremum* of a set A of ordinals is $\bigcup A$. The usual *order relations* are introduced on ORD:

$$\alpha < \beta \leftrightarrow \alpha \in \beta \quad \alpha \leq \beta \leftrightarrow \alpha \subseteq \beta.$$

It is now easily shown that successors and suprema of ordinals are again ordinals and that

$$\alpha < \beta \leftrightarrow \alpha^+ \leq \beta \quad \bigcup A \leq \beta \leftrightarrow \forall \alpha \in A\, \alpha < \beta \leq \gamma \to \alpha < \gamma.$$

But straightforward arguments show that any of the following assertions (for arbitrary ordinals α, β, γ) implies LEM: (i) $\alpha < \beta \vee \alpha = \beta \vee \beta < \alpha$, (ii) $\alpha \leq \beta \vee \beta \leq \alpha$, (iii) $\alpha \leq \beta \to \alpha < \beta \vee \alpha = \beta$, (iv) $\alpha < \beta \to \alpha^+ < \beta \vee \alpha^+ = \beta$, (v) $\alpha \leq \beta < \gamma \to \alpha < \gamma$.

Definition   An ordinal α is a *successor* if $\exists \beta\ \alpha = \beta^+$, a *weak limit* if $\forall \beta \in \alpha\, \exists\, \gamma \in \alpha\, \beta \in \gamma$, and a *strong limit* if $\forall \beta \in \alpha\, \beta^+ \in \alpha$.

Note that both the following assertions imply LEM: (i) every ordinal is zero, a successor, or a weak limit, (ii) all weak limits are strong limits. Assertion (i) follows from the observation that, for any formula φ, if the specified disjunction applies to the ordinal $\{0|\varphi\}$, then $\varphi \vee \neg\varphi$. As for assertion (ii), define

$$1_\varphi = \{0|\varphi\}, 2_\varphi = \{0, 1_\varphi\}, \beta = \{0, 1_\varphi, 2_\varphi, 2_\varphi{}^+, 2_\varphi{}^{++}, \ldots\}.$$

Then β is a weak limit, but a strong one only if $\varphi \vee \neg\varphi$.

As in classical set theory, in IZF a connection can be established between the class of ordinals and certain natural notions of well-founded or well-ordered structure. Thus a *well-founded* relation on a set A is a binary relation which is *inductive*, that is,

$$\forall X \subseteq A[\forall x \in A[\forall y < x \cdot y \in X \to x \in X] \to A \subseteq X].$$

A well-founded relation has no infinite descending sequences and so is irreflexive. Moreover, the usual proof may be given in IZF to justify *definitions by recursion* on a well-founded relation, so that we can make the

> **Definition** If $<$ is a well-founded relation on a set A, the associated *rank function* $\rho_<\colon A \to \mathrm{ORD}$ is the (unique) function such that for each $x \in A$,
>
> $$\rho_<(x) = \bigcup\{\rho_<(y)^+\colon y < x\}.$$

When $<$ is $\in$ restricted to an ordinal, it is easy to see that the associated rank function is the identity.

To obtain a characterization of the *order-types* represented by ordinals we make the

> **Definition** A binary relation $<$ on a set A is *transitive* if $\forall x \in A \forall y \in A \forall z \in A(x < y \wedge y < z \to x < z)$, and *extensional* if $\forall x \in A \forall y \in A[\forall z(z < x \leftrightarrow z < y) \to x = y]$. A *well-ordering* is a transitive, extensional well-founded relation.

It is then easily shown that the well-orderings are exactly those relations isomorphic to $\in$ restricted to some ordinal. For it follows immediately from the axioms of regularity and extensionality that the $\in$-relation well-orders every ordinal. And conversely, it is easy to prove by induction that the rank assigning function on any well-ordering is an isomorphism.

Heyting-algebra-valued models

We now argue in ZFC. Suppose given a complete Heyting algebra H. We obtain the H-*valued universe*[4] $V^{(H)}$ by carrying out the definition of the Boolean-valued universe $V^{(B)}$ with H in place of B; for sentences σ of $\mathcal{L}^{(H)}$—the language obtained from $\mathcal{L}$ by adding a name for each member of $V^{(H)}$—the H-value $[\![\sigma]\!]^H$ (written simply $[\![\sigma]\!]$) of σ in $V^{(H)}$ is defined analogously. Many of the results established in Chapter 1 for $V^{(B)}$ hold, *mutatis mutandis*, for $V^{(H)}$, and are proved in an analogous way: these include 1.7, Corollary 1.18, Theorem 1.23 (based on Definition 1.22 of the canonical embedding $\hat{\ }$, now from V into $V^{(H)}$), Lemma 1.25, Problem 1.29, and Lemma 1.31. Theorem 1.17 now takes the form:

> *In $V^{(H)}$, all the axioms of intuitionistic first-order logic are true, and all its rules of inference are valid. Clauses (i)–(vii) also continue to hold.*

Theorem 1.33 now reads:

> *All the axioms of IZF are true in $V^{(H)}$.*

---

[4]Although the definition of $V^{(H)}$ can carried out, and its properties established, in IZF, this is more easily done in ZFC.

Clearly LEM holds $V^{(H)}$ if and only if H is a Boolean algebra. Since AC implies LEM, the former does not hold in any $V^{(H)}$ for which H is not a Boolean algebra (for example, the algebra of opens of the space of real numbers). But, interestingly, *Zorn's lemma* is *always* valid in $V^{(H)}$. Here is a sketch of the argument, in which Zorn's lemma is taken in the form: *any nonempty partially ordered set in which every chain has a supremum also has a maximal element.* Thus suppose $X, \leq_X \in V^{(H)}$ satisfy

$$V^{(H)} \models \ \langle X, \leq_X \rangle \ \textit{is a nonempty partially ordered set in which}$$
$$\textit{every chain has a supremum.}$$

Let Y be a core for X and define the relation $\leq_Y$ on Y by $y \leq_Y y' \leftrightarrow [\![y \leq_X y']\!] = 1$; it is then easily verified that $\langle Y, \leq_Y \rangle$ is a partially ordered set in which every chain has a supremum. So, by Zorn's lemma in V, Y has a maximal element c. We claim that

$$[\![c \textit{ is a maximal element of } X]\!] = 1. \tag{1}$$

To prove (1) we take any $a \in V^{(H)}$ and define $V \in V^{(H)}$ by $\mathrm{dom}(V) = \mathrm{dom}(X)$ and

$$V(x) = [\![x = a \wedge x \in X \wedge c \leq_X x]\!] \vee [\![x = c]\!].$$

For $x \in \mathrm{dom}(X)$. It is then readily verified that $V^{(H)} \models V$ *is a chain in* X; and so (using Problem 1.29 for $V^{(H)}$) there is $v \in Y$ for which

$$V^{(H)} \models v \textit{ is the supremum of } V. \tag{2}$$

Since $[\![c \in V]\!] = 1$ it follows that $[\![c \leq_X v]\!] = 1$, whence $c \leq_Y v$, so that $v = c$ by the maximality of c. This and (2) now yield $[\![a \in V \to a \leq_X c]\!] = 1$; and clearly $[\![a \in V \to c \leq_X x]\!] = 1$. Therefore

$$[\![a \in V \to a = c]\!] = 1. \tag{3}$$

It is easily verified that

$$[\![a \in X \wedge c \leq_X a]\!] \leq [\![a \in V]\!]. \tag{4}$$

(3) and (4) yield $[\![a \in X \wedge c \leq_X a]\!] \leq [\![a = c]\!]$; since this holds for arbitrary $a \in V^{(H)}$, (1) follows.

From the fact that Zorn's lemma holds in every $V^{(H)}$ but AC does not we may infer that, in IZF, the former does not imply the latter. In IZF Zorn's lemma is thus very weak, indeed so weak as to be entirely compatible with intuitionistic logic. For more on this see Bell (1997).

Forcing in Heyting-algebra-valued models and independence in IZF

Let $\langle P, \leq \rangle$ be a partially ordered set and for each $p \in P$ define $O_p = \{q \in P : q \leq p\}$. Assign P the topology with base $\{O_p : p \in P\}$ and write H for the complete Heyting algebra $O(P)$ of open sets in P. It is easy to see that the members of $O(P)$ are precisely the *downward-closed* subsets of P, that is, subsets U of P satisfying $x \in U$ and $y \leq x \to y \in U$. The map $p \mapsto O_p$ is clearly order preserving from P to H; let us agree to identify p with O_p for each $p \in P$, thus identifying P as a subset of H.

We define the *intuitionistic forcing relation* $\Vdash$ between elements of P (which will now be known as *forcing* conditions) and sentences of $\mathcal{L}^{(H)}$ by

$$p \Vdash \sigma \text{ iff } p \leq [\![\sigma]\!].$$

Intuitionistic forcing satisfies the following conditions:

- $p \Vdash \sigma \wedge \tau \leftrightarrow p \Vdash \sigma \ \& \ p \Vdash \tau$
- $p \Vdash \sigma \vee \tau \leftrightarrow p \Vdash \sigma \text{ or } p \Vdash \tau$
- $p \Vdash \sigma \to \tau \leftrightarrow \forall q \leq p[q \Vdash \sigma \to q \Vdash \tau]$
- $p \Vdash \neg \sigma \leftrightarrow \forall q \leq p \text{ not } q \Vdash \varphi$
- $p \Vdash \forall x \varphi \leftrightarrow p \Vdash \varphi(a)$ for every $a \in V^{(H)}$
- $p \Vdash \exists x \varphi \leftrightarrow p \Vdash \varphi(a)$ for some $a \in V^{(H)}$.

Cohen forcing $\Vdash_c$ (mentioned in Chapter 2) differs from intuitionistic forcing only in that the clause for $\forall$ is replaced by

$$p \Vdash_c \forall x \varphi \leftrightarrow \forall q \leq p\, q \nVdash_c \varphi(a) \text{ for every } a \in V^{(H)},$$

or equivalently

$$p \Vdash_c \forall x \varphi \leftrightarrow p \Vdash_c \neg \exists x \neg \varphi.$$

Clearly, $V^{(H)} \models \sigma$ iff $p \Vdash \sigma$ for all p, that is, truth in $V^{(H)}$ coincides with the property of being universally forced. This explains the fact, puzzling at the time of its introduction, that Cohen forcing satisfies intuitionistic rather than classical rules: it reflects truth in the intuitionistic model $V^{(O(P))}$, rather than in the Boolean-valued model $V^{(\mathrm{RO}(P))}$.

In order to establish the independence of the axiom of choice from ZF we constructed in Chapter 3 a Boolean-valued model containing an infinite but Dedekind finite set. As we shall see, it is a much simpler task to construct a Heyting-algebra-valued model of IZF with the same property.

Define the set $K \in V^{(H)}$ by $\mathrm{dom}(K) = \{\hat{p} : p \in P\}$ and $K(\hat{p}) = O_p$. Then, in $V^{(H)}$, K is a subset of $\hat{P}$ and for $p \in P$, $[\![\hat{p} \in K]\!] = O_p$. Now consider the case in which P is the opposite $\mathbb{N}^{op}$ of the totally ordered set $\mathbb{N}$ of natural numbers.

Here the associated complete Heyting algebra $H = O(\mathbb{N}^{op})$ is the family of all *upward*-closed sets of natural numbers. In this case, as we shall see, the H-valued set K *is infinite and Dedekind finite in* $V^{(H)}$.

To see this, first note that $V^{(H)} \models K \subseteq \hat{\mathbb{N}}$ and that $V^{(H)} \models \neg\neg\forall n \in \hat{\mathbb{N}}\, n \in K$. But then, working in $V^{(H)}$, if $\forall n \in \hat{\mathbb{N}}\, n \in K$, then K is not finite, so if K were finite, then $\neg\forall n \in \hat{\mathbb{N}}\, n \in K$, and so $\neg\neg\forall n \in \hat{\mathbb{N}}\, n \in K$ yields the nonfiniteness of K.

But, in $V^{(H)}$, K is Dedekind finite. For (again working in $V^{(H)}$), if K were Dedekind infinite (i.e. if there existed an injection of $\hat{\mathbb{N}}$ into K), then the sentence $\forall x \in K \exists y \in K\, x < y$ would also have to hold in $V^{(H)}$. But calculating the truth value of that sentence gives:

$$
\begin{aligned}
\llbracket \forall x \in K \exists y \in K\, x < y \rrbracket &= \bigcap_{m \in \mathbb{N}^{op}} [O_m \Rightarrow \bigcup_{n \in \mathbb{N}^{op}} [O_n \cap \llbracket \hat{m} < \hat{n} \rrbracket]] \\
&= \bigcap_m [O_m \Rightarrow \bigcup_{m<n} O_n] \\
&= \bigcap_m [O_m \Rightarrow O_{m+1}] \\
&= \bigcap_m O_{m+1} \\
&= \emptyset
\end{aligned}
$$

Therefore $\forall x \in K \exists y \in K\, x < y$ is false in $V^{(H)}$ and so K is not Dedekind infinite.

Many other independence results from IZF can be established by the use of Heyting-algebra-valued models. For example, in Fourman and Hyland (1979) such models are presented in which

- the sets of Cauchy and Dedekind real numbers fail to coincide
- the field of complex numbers fails to be algebraically closed
- every function from $\mathbb{R}$ to $\mathbb{R}$ is continuous.

APPENDIX

BOOLEAN AND HEYTING ALGEBRA-VALUED MODELS AS CATEGORIES

In this appendix we provide a brief introduction to category theory[1] and describe how a Boolean and Heyting algebra-valued model can be viewed as a category of a particularly significant kind known as a *topos*.

Categories and functors

A *category* $\mathscr{C}$ is determined by first specifying two classes $Ob(\mathscr{C})$, $Arr(\mathscr{C})$—the collections of $\mathscr{C}$-*objects* and $\mathscr{C}$-*arrows*. These collections are subject to the following axioms:

- Each $\mathscr{C}$-arrow f is assigned a pair of $\mathscr{C}$-objects $\mathrm{dom}(f)$, $\mathrm{cod}(f)$ called the *domain* and *codomain* of f, respectively. To indicate the fact that $\mathscr{C}$-objects X and Y are respectively the domain and codomain of f we write $f\colon X \to Y$ or $X \xrightarrow{f} Y$. The collection of $\mathscr{C}$-arrows with domain X and codomain Y is written $\mathscr{C}(X, Y)$.
- Each $\mathscr{C}$-object X is assigned a $\mathscr{C}$-arrow $1_X\colon X \to X$ called the *identity arrow* on X.
- Each pair f, g of $\mathscr{C}$-arrows such that $\mathrm{cod}(f) = \mathrm{dom}(g)$ is assigned an arrow $g \circ f\colon \mathrm{dom}(f) \to \mathrm{cod}(g)$ called the *composite* of f and g. Thus if $f\colon X \to Y$ and $g\colon Y \to Z$ then $g \circ f\colon X \to Z$. We also write $X \xrightarrow{f} Y \xrightarrow{g} Z$ for $g \circ f$. Arrows f, g satisfying $\mathrm{cod}(f) = \mathrm{dom}(g)$ are called *composable*.
- *Associativity law.* For composable arrows (f, g) and (g, h), we have $h \circ (g \circ f) = (h \circ g) \circ f$.
- *Identity law.* For any arrow $f\colon X \to Y$, we have $f \circ 1_X = f = 1_Y \circ f$.

As a basic example of a category, we have the category Set of sets whose objects are all sets and whose arrows are all maps between sets (strictly, triples (f, A, B) with $\mathrm{domain}(f) = A$ and $\mathrm{range}\ (f) \subseteq B$.) Other examples of categories are the category of groups, with objects all groups and arrows all group homomorphisms and the category of topological spaces with objects all topological spaces and arrows all continuous maps. Categories with just one object may be identified with *monoids*, that is, algebraic structures with an associative multiplication and an identity element.

---

[1] Useful accounts of category theory include Mac Lane (1971) and McLarty (1992).

We list some basic category-theoretic definitions:

<table>
<tr><td>

Commutative diagram
(in category)

</td><td>

Diagram of objects and arrows such that the arrow obtained by composing the arrows of any connected path depends only on the endpoints of the path.

</td></tr>
<tr><td>

Initial object

</td><td>

Object 0 such that, for any object X, there is a unique arrow $0 \to X$ (e.g. $\varnothing$ in Set)

</td></tr>
<tr><td>

Terminal object

</td><td>

Object 1 such that, for any object X, there is a unique arrow $X \to 1$ (e.g. $\{0\}$ in Set)

</td></tr>
<tr><td>

Element of an object X

</td><td>

Arrow $1 \to X$ (in Set, member of X)

</td></tr>
<tr><td>

Monic arrow $X \rightarrowtail Y$

</td><td>

Arrow $f\colon X \to Y$ such that, for any arrows $g, h\colon Z \to X$, $f \circ g = f \circ h \Rightarrow g = h$ (in Set, *one–one* map)

</td></tr>
<tr><td>

Epic arrow $X \twoheadrightarrow Y$

</td><td>

Arrow $f\colon X \to Y$ such that, for any arrows $g, h\colon Y \to Z, g \circ f = h \circ f \Rightarrow g = h$ (in Set, *onto* map)

</td></tr>
<tr><td>

Isomorphism $X \cong Y$

</td><td>

Arrow $f\colon X \to Y$ for which there is $g\colon Y \to X$ such that $g \circ f = 1_X, f \circ g = 1_Y$ (in Set, *bijection*)

</td></tr>
<tr><td>

Decomposition of arrow
$f\colon A \to B$

</td><td>

Pair of arrows with $A \xrightarrow{k} C \xrightarrow{m} B$ with k epic, m monic and $m \circ k = f$. The decomposition (k, m) of f is said to be *unique* (up to isomorphism if for any decomposition $A \xrightarrow{k'} C' \xrightarrow{m'} B$ of f there is an isomorphism $i\colon C \cong C'$ such that the diagram

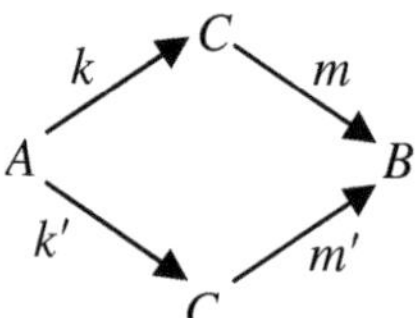

commutes. (In Set, the decomposition of a map $f\colon A \to B$ is the pair consisting of f considered as a surjection onto $f[A]$ and the insertion map of $f[A]$ into B.)

</td></tr>
<tr><td>

Image of arrow
$f\colon A \to B$

</td><td>

Monic m in a unique decomposition (k, m) of f

</td></tr>
<tr><td>

Amenable category

</td><td>

Category in which every arrow admits a unique decomposition

</td></tr>
</table>

Product of objects
X, Y

Object $X \times Y$ with arrows (projections)
$X \xleftarrow{\pi_1} X \times Y \xrightarrow{\pi_2} Y$ such that any diagram

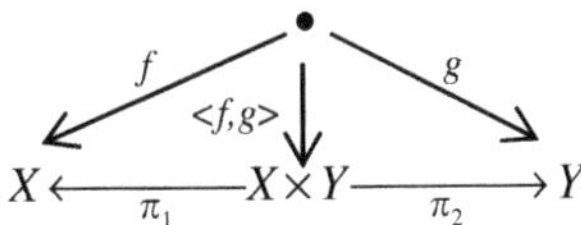 can be uniquely completed to a
commutative diagram

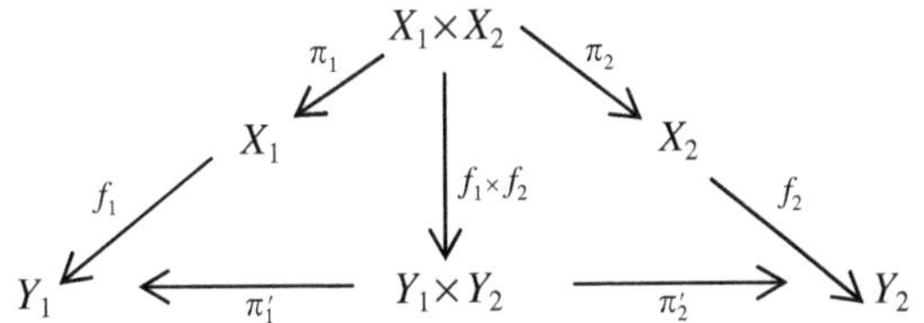

(In Set, Cartesian product of sets with projection
maps.)

Product of arrows
$f_1 \colon X_1 \to Y_1$
$f_2 \colon X_2 \to Y_2$

Unique arrow $f_1 \times f_2 \colon X_1 \times X_2 \to Y_1 \times Y_2$ making
the diagram

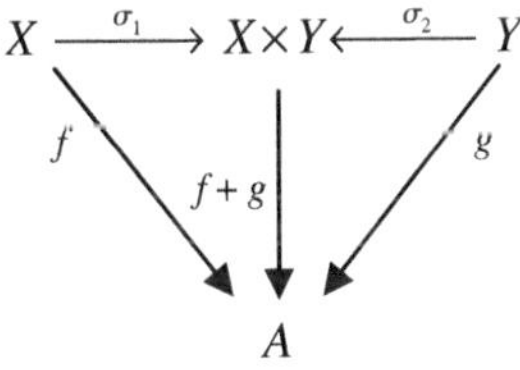

commute. That is, $f_1 \times f_2 = \langle f_1 \circ \pi_1, \ f_2 \circ \pi_2 \rangle$.

Coproduct of objects
X, Y

Object $X + Y$ together with a pair of arrows
$X \xrightarrow{\sigma_1} X + Y \xleftarrow{\sigma_2} Y$ such that for any pair of
arrows $X \xrightarrow{f} A \xleftarrow{g} B$, there is a unique arrow
$X + Y \xrightarrow{f+g} A$ such that the diagram

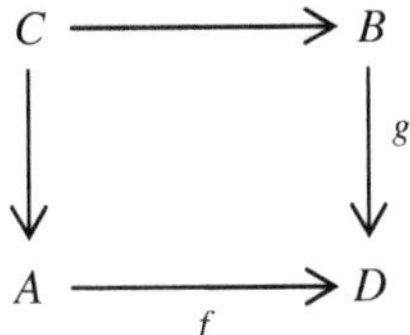

commutes. (In Set, disjoint union with insertions
maps.)

Pullback diagram or
square

Commutative diagram of the form

such that for any commutative diagram

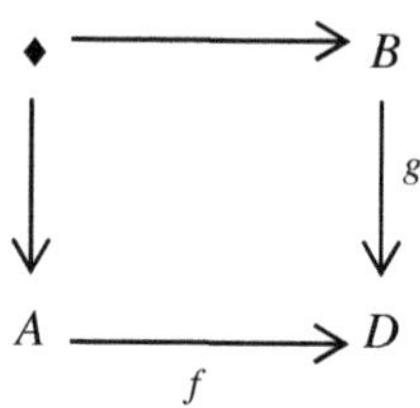

there is a unique $\blacklozenge \xrightarrow{\;!\;} C$ such that

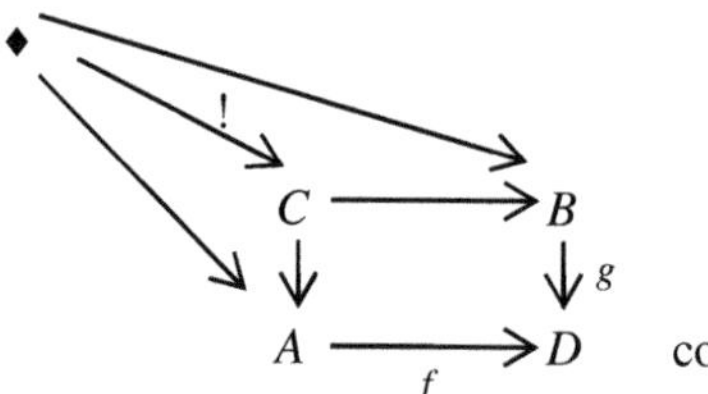   commutes.

(In Set, C is $\{\langle x,\, y\rangle\colon f(x) = g(y)\}$. If $D{=}1$, then C is $A \times B$; if $A \subseteq D$ and f is the insertion map, then C is $g^{-1}[A]$; if $A,\, B \subseteq D$ and f and g are insertions, then C is $A \cap B$.

Equalizer of pair of arrows

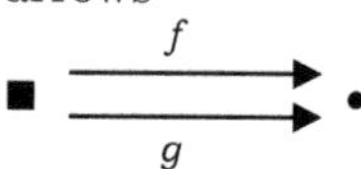

Arrow $\blacklozenge \xrightarrow{\;e\;} \blacksquare$ such that $f \circ e = g \circ e$ and, for any arrow $\blacktriangle \xrightarrow{\;e'\;} \blacksquare$ such that $f \circ e' = g \circ e'$ there is a unique $\blacktriangle \xrightarrow{\;u\;} \blacklozenge$ such that

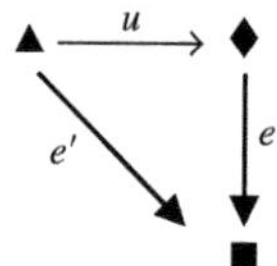

commutes. (In Set, $\{x\colon f(x) = g(x)\}$ with insertion map.)

Subobject of an object X   Pair $(m,\, Y)$, with m a monic arrow $Y \rightarrowtail X$

Truth value object or *subobject classifier*   Object Ω together with arrow *true*: $1 \to \Omega$ such that every monic $m\colon \bullet \rightarrowtail \blacklozenge$ (i.e. subobject of $\blacklozenge$) can be uniquely extended to a pullback diagram of the form

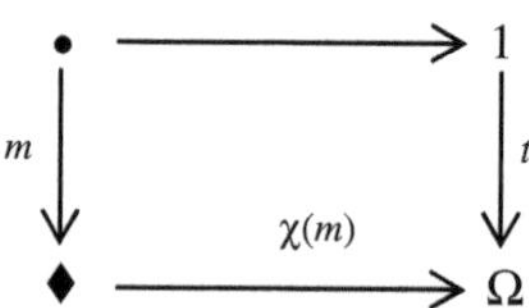

$\chi(m)$ is the *characteristic arrow* of m. (In Set, Ω is the set $2 = \{0,\ 1\}$ t is 1 and $\chi(m)$ is the characteristic function of the image of m.) For any object A, $\chi(1_A)$ is written T_A.

Power object of an object X.

An object PX together with an arrow (evaluation) $e_X\colon X \times PX \to \Omega$ such that, for any $f\colon X \times PX \to \Omega$, there is a unique arrow $Y \to PX$ such that

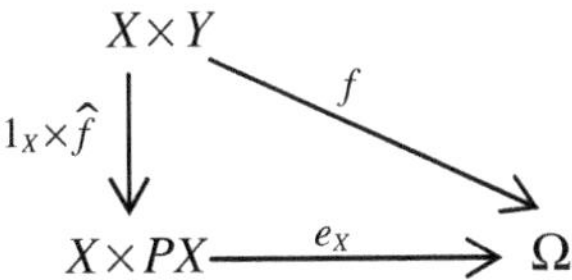

commutes. (In Set, PX is the power set of X and e_X the characteristic function of the membership relation between X and PX.)

Exponential object of objects $Y,\ X$

An object Y^X, together with an arrow $ev\colon X \times Y^X \to Y$ such that, for any arrow $f\colon X \times Z \to Y$ there is a unique arrow $\hat{f}\colon Z \to Y^X$—the *exponential transpose of* f—such that the diagram

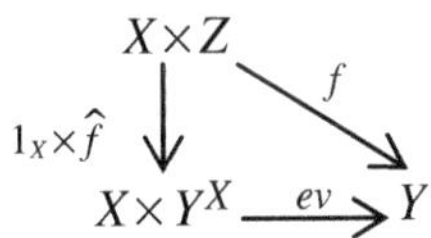

commutes. In Set, Y^X is the set of all maps $X \to Y$ and ev is the map that sends (x, f) to $f(x)$.

Product of indexed set $\{A_i\colon i \in I\}$ *of objects*

Object $\prod_{i \in I} A_i$ together with arrows $\prod_{i \in I} A_i \xrightarrow{\pi_i} A_i$ $(i \in I)$ such that, for any arrows $f_i\colon B \to A_i$ $(i \in I)$ there is a *unique* arrow $h\colon B \longrightarrow \prod_{i \in I} A_i$ such that, for each $i \in I$, the diagram

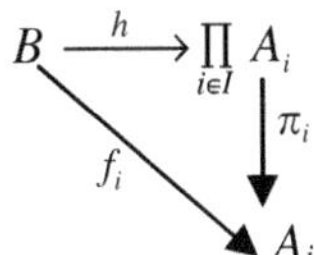

commutes.

Coproduct of indexed set $\{A_i\colon i \in I\}$ of objects	Objects $\coprod_{i\in I} A_i$ together with arrows $A_i \xrightarrow{\sigma_i} \coprod_{i\in I} A_i$ $(i \in I)$ such that, for any arrows $f_i\colon A_i \to B$ $(i \in I)$ there is a *unique* arrow $h\colon \coprod_{i\in I} A_i \longrightarrow B$ such that, for each $i \in I$, the diagram 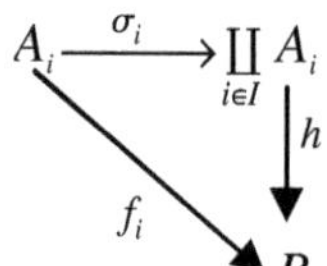 commutes.

A category is said to satisfy the *axiom of choice* if for any epic $f\colon A \twoheadrightarrow B$ there is a (necessarily monic) $g\colon B \to A$ such that $f \circ g = 1_B$. We assume that the axiom of choice holds in **Set**: this is equivalent to assuming the axiom of choice in any of its usual forms.

A *functor* $F\colon \mathscr{C} \to \mathscr{D}$ between two categories $\mathscr{C}$ and $\mathscr{D}$ is a map that 'preserves commutative diagrams', that is, assigns to each $\mathscr{C}$-object A a $\mathscr{D}$-object FA and to each $\mathscr{C}$-arrow $f\colon A \to B$ a $\mathscr{D}$-arrow $Ff\colon FA \to FB$ in such a way that:

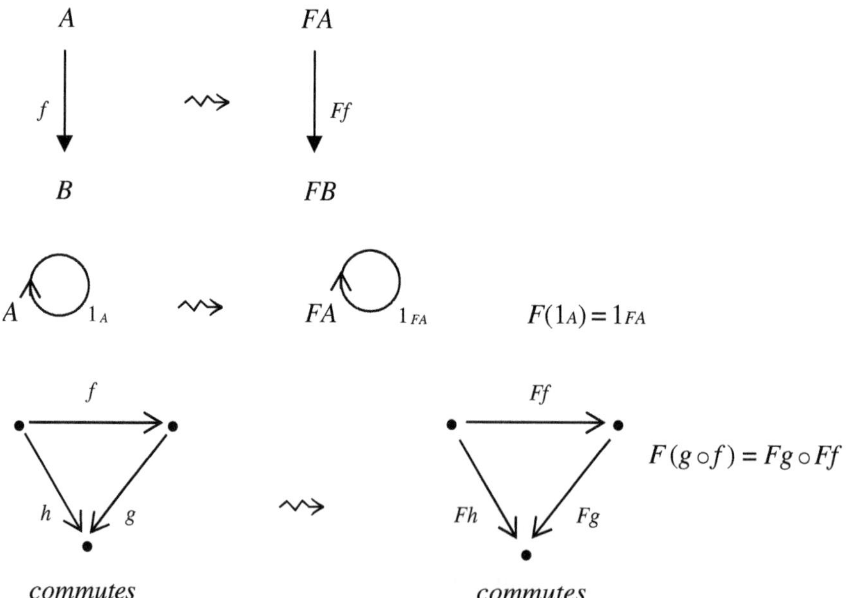

Two functors $F\colon \mathscr{C} \to \mathscr{D}$ and $G\colon \mathscr{D} \to \mathscr{E}$ can obviously be composed to yield a functor $GF\colon \mathscr{C} \to \mathscr{E}$. Associated with each category $\mathscr{C}$ we have the evident *identity functor* $1_{\mathscr{C}}\colon \mathscr{C} \to \mathscr{C}$.

Given functors F, $G\colon \mathscr{C} \to \mathscr{D}$, a *natural transformation* between F and G is a map η from the objects of $\mathscr{C}$ to the arrows of $\mathscr{D}$ satisfying the following conditions.

- For each object A of $\mathscr{C}$, ηA is an arrow $FA \to GA$ in $\mathscr{D}$
- For each arrow $f\colon A \to A'$ in $\mathscr{C}$, the diagram

$$
\begin{array}{ccc}
FA & \xrightarrow{\ \eta A\ } & GA \\
{\scriptstyle Ff}\downarrow & & \downarrow{\scriptstyle Gf} \\
FA' & \xrightarrow{\ \eta A'\ } & GA'
\end{array}
$$

commutes.

When each ηA is an isomorphism we say that η is a *natural isomorphism* between F and G and that F and G are *naturally isomorphic*, written $F \cong G$.

A functor $F\colon \mathscr{C} \to \mathscr{D}$ is an *equivalence* if it is 'an isomorphism up to isomorphism', that is, if it is

- *faithful*: $Ff = Fg \Rightarrow f = g$.
- *full*: for any $h\colon FA \to FB$ there is $f\colon A \to B$ such that $h = Ff$.
- *dense*: for any $\mathscr{D}$-object B there is a $\mathscr{C}$-object A such that $B \cong FA$.

An equivalence $F\colon \mathscr{C} \to \mathscr{D}$ can also be characterized as a functor admitting a *quasi-inverse*, that is, for which there exists a functor $G\colon \mathscr{D} \to \mathscr{C}$ such that FG and GF are naturally isomorphic to the identity functors on $\mathscr{D}$ and $\mathscr{C}$ respectively. Two categories are *equivalent*, written $\simeq$, if there is an equivalence between them. Equivalence is the appropriate notion of 'identity of form' for categories.

Toposes

A category is *cartesian closed* if it has a terminal object, as well as products and exponentials of arbitrary pairs of its objects. It is *finitely complete* if it has a terminal object, products of arbitrary pairs of its objects, and equalizers. An (elementary) *topos*[2] is a category possessing a terminal object, products, a truth-value object, and power objects. It can be shown that every topos is

---

[2] Accounts of topos theory may be found in Bell (1988), Goldblatt (1979), Johnstone (1977) and (2002), Lambek and Scott (1986), Mac Lane and Moerdijk (1992), and McLarty (1992). See also Mac Lane (1975).

cartesian closed, finitely complete, and amenable. The basic example of a topos is Set: its terminal and truth-value objects, products and power objects have been identified above.

Here are some further examples of toposes.

Set$^\rightarrow$: topos of sets varying over two possible states 0 (then), 1 (now), with $0 \leq 1$. An *object* X here is a pair of sets X_0, X_1 together with a 'transition' map $p\colon X_0 \to X_1$. An *arrow* $F\colon X \to Y$ is a pair of maps $f_0\colon X_0 \to Y_0, f_1\colon X_1 \to Y_1$ compatible with the transition maps in the sense that the diagram

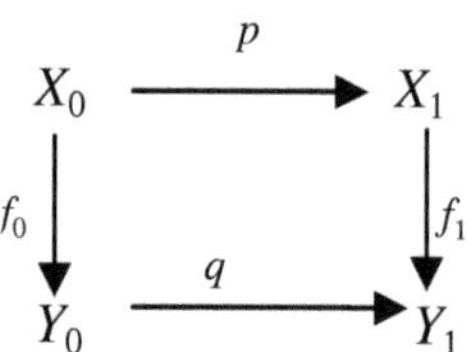

commutes.

Notice that the truth value object Ω in Set$^\rightarrow$ has 3 (rather than 2) elements. For if (m, X) is a subobject of Y in Set$^\rightarrow$, then we may take $X_0 \subseteq Y_0$, $X_1 \subseteq Y_1$, f_0 and f_1 identity maps, and p to be the restriction of q to X_0. Then for any $y \in Y$ there are three possibilities, as depicted below: (0) $q(y) \in X_1$ and $y \notin X_0$, (1) $y \in X_0$, and (2) $q(y) \notin X_1$.

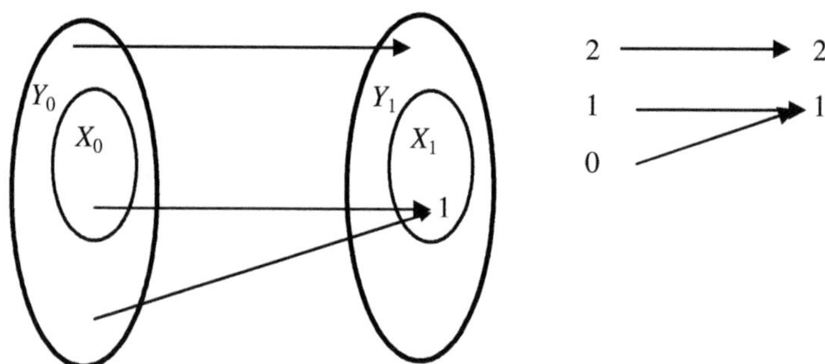

So if $2 = \{0, 1\}$ and $3 = \{0, 1, 2\}$ we take Ω to be the variable set $3 \to 2$ with $0 \mapsto 1$, $1 \mapsto 1$, $2 \mapsto 2$.

More generally, we may consider sets varying over n, or ω, or any totally ordered 'number' of stages. In each case there is 'one more' truth value than stages: 'truth' = 'time' + 1.

Still more generally, we may consider the category Set^P of *sets varying over partially ordered set P*. As objects this category has functors $P \to$ Set, that is, maps F which assign to each $p \in P$ a set $F(p)$ and to each $p, q \in P$ such that

$p \leq q$ a map $F_{pq}\colon F(p) \to F(q)$ satisfying:

$p \leq q \leq r$ implies that

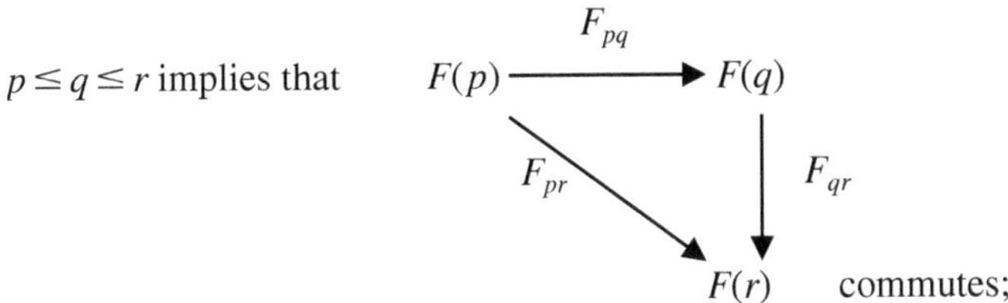

$F(r)$ commutes;

and

$$F_{pp} \text{ is the identity map on } F(p).$$

An *arrow* $\eta\colon F \to G$ in Set^P is a natural transformation between F and G, in this case, an assignment of a map $\eta_p\colon F(p) \to G(p)$ to each $p \in P$ in such a way that, whenever $p \leq q$, the diagram

$$
\begin{array}{ccc}
F(p) & \xrightarrow{\ F_{pq}\ } & F(q) \\
\downarrow{\scriptstyle \eta_p} & & \downarrow{\scriptstyle \eta_q} \\
G(p) & \xrightarrow[\ G_{pq}\ ]{} & G(q)
\end{array}
\quad \text{commutes.}
$$

The truth-value object Ω in Set^P is determined as follows. A subset U of $O_p = \{q \in P : p \leq q\}$ such that $q \in U, r \geq q \Rightarrow r \in U$ is said to be *upward closed* over p. Then

$$\Omega(p) = \text{ family of all upward-closed sets over } p,$$
$$\Omega_{pq}(U) = U \cap O_q \text{ for } p \leq q, \ U \in \Omega(p).$$

The *terminal object* 1 in Set^P is the functor on P with constant value $1 = \{0\}$ and *true*$\colon 1 \to \Omega$ has $\mathit{true}_p(0) = O_p$ for each $p \in P$.

Objects in $\mathsf{Set}^{P^{op}}$, where P^{op} is the partially ordered set obtained by reversing the order on P, are called *presheaves* on P. If F is a presheaf on P, $x \in F(p)$, and $q \leq p$, we write $x \upharpoonright_F q$ for $F_{pq}(x)$.

Now let H be a complete Heyting algebra. A presheaf F on H is a *sheaf* if whenever $p = \bigvee_{i \in I} p_i$ in H and $s_i \in F(p_i)$ for all $i \in I$ satisfy $s_i \upharpoonright_F (p_i \cap p_i) = s_j \upharpoonright_F (p_i \cap p_j)$ for all $i, j \in I$, then there is a *unique* $s \in F(U)$ such that $s \upharpoonright_F p_i = s_i$ for all $i \in I$. The category $\mathsf{Shv}(H)$ of *sheaves on H* has as objects the sheaves on

H and as arrows and as arrows all arrows between these objects *qua* presheaves. It can be shown that $\mathsf{Shv}(H)$ is a topos.

In any topos $\mathscr{E}$ the truth-value object Ω supports natural arrows representing the familiar logical operations $\wedge,\ \vee,\ \Rightarrow,\ \neg$. To wit,

$\wedge : \Omega \times \Omega \to \Omega$ is the characteristic arrow of the monic

$<true,\ true> : 1 \to \Omega \times \Omega;$

$\vee : \Omega \times \Omega \to \Omega$ is the characteristic arrow of the

image of $\Omega + \Omega \xrightarrow{\ <T_\Omega,1_\Omega>+<1_\Omega,T_\Omega>\ } \Omega + \Omega;$

$\neg : \Omega \to \Omega$ is the characteristic arrow of $false : 1 \to \Omega$, where *false* is the

characteristic arrow of the monic $0 \to 1;$

$\Rightarrow \Omega \times \Omega \to \Omega$ is the characteristic arrow of the equalizer of the pair

of arrows $\pi_1, \wedge \colon \Omega \times \Omega \to \Omega$.

It can then be shown that with these operations Ω is an 'internal Heyting algebra' in $\mathscr{E}$ in the sense the diagrams in $\mathscr{E}$ representing equations characterizing Heyting algebras (see Ch. 0) all commute. For example, the commutative diagram corresponding to the equation $x \wedge (x \Rightarrow y)$ is

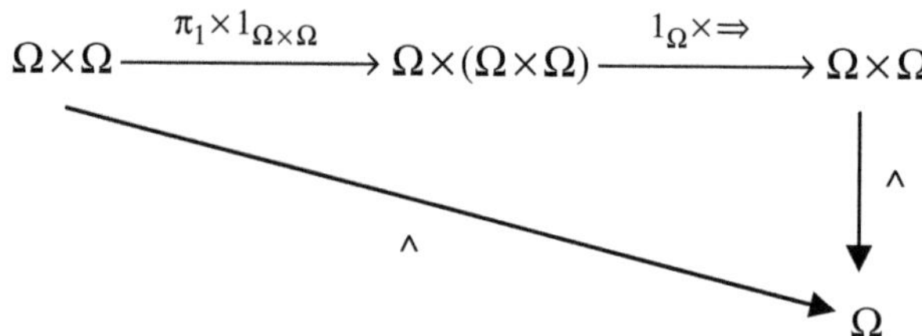

Ω is an internal Boolean algebra in $\mathscr{E}$ if $\neg \circ \neg = 1_\Omega$. A topos satisfying this condition is called *Boolean*; clearly the topos Set is Boolean. It is not difficult to show that each of the following conditions on a topos $\mathscr{E}$ are necessary and sufficient for it to be Boolean: (i) $1 + 1 \xrightarrow{\ true+false\ } \Omega$ is an isomorphism; (ii) all subobjects in $\mathscr{E}$ have *complements*, that is, given a monic $U \rightarrowtail X$ there is a monic $V \rightarrowtail X$ such that $U + V \to X$ is an isomorphism.

Diaconescu's theorem asserts that any topos in which the axiom of choice holds is Boolean. This is the category-theoretic form of the fact (proved in Ch. 8) that in intuitionistic set theory the axiom of choice implies the law of excluded middle.

Recall that an *element* of an object X is an arrow $1 \to X$. If $\mathscr{E}$ is a topos, the set $\Omega(\mathscr{E})$ of elements of Ω in $\mathscr{E}$ can be assigned a partial ordering $\leq$ by defining $u \leq v$ iff $u \wedge v = u$; more exactly, if the diagram

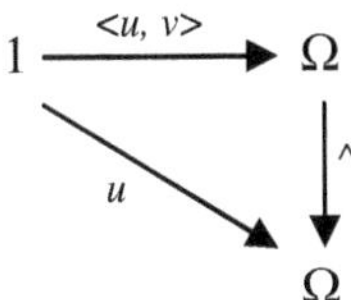

commutes. From the fact that Ω is an internal Heyting algebra it follows that, with this partial ordering, $\Omega(\mathscr{E})$ is a Heyting algebra, the *algebra of external truth values of $\mathscr{E}$*. When $\mathscr{E}$ is Boolean, $\Omega(\mathscr{E})$ is a Boolean algebra.[3] A topos is *bivalent* if its algebra of external truth values contains just the two elements *true* and *false*. Clearly Set is bivalent.

$\Omega(\mathscr{E})$ is *complete* when $\mathscr{E}$ admits arbitrary coproducts of 1, that is, if for any set I the I-indexed coproduct $\coprod_I 1$ exists in $\mathscr{E}$. Although not every topos satisfies this condition, it is satisfied by all the toposes we have mentioned. Here is a table giving $\Omega(\mathscr{E})$ for various toposes $\mathscr{E}$.

$\mathscr{E}$	$\Omega(\mathscr{E})$
Set	2
Set$^{P^{op}}$	$O(P)$ with P assigned order topology
Shv(H)	H

Boolean and Heyting algebra-valued-models as toposes

If H is a complete Heyting algebra, the 'topos of sets constructed within $V^{(H)}$' is also a topos. More precisely, we obtain a topos Set$^{(H)}$ from $V^{(H)}$ in the following way. First, we identify elements u, v of $V^{(H)}$ when $[\![u = v]\!] = 1$. The objects of the category Set$^{(H)}$ are the (thus identified) objects of $V^{(H)}$ and the arrows of Set$^{(H)}$ are those (identified) objects f of $V^{(H)}$ for which $[\![f \text{ is a function}]\!] = 1$. Composition and identity arrows are defined in the obvious way. Now it is easy to derive in IZF the assertion that Set is a topos with subject classifier $P1$. So since the axioms of IZF are all true in $V^{(H)}$, it follows that Set$^{(H)}$ *is a topos with subobject classifier* $\Omega = P^{(H)}(\hat{1})$.[4] For each set I, the H-valued set $\hat{I}$ is readily shown to be the I-indexed copower of 1 in Set$^{(H)}$. Elements of Ω in Set$^{(H)}$ correspond to the $u \in V^{(H)}$ for which $[\![u \in \Omega]\!] = [\![u \subseteq \hat{1}]\!] = 1_H$. These are in turn correlated with the elements of H via the map $u \longmapsto [\![\hat{0} \in u]\!]$. Thus the

---

[3] The converse, however, does not hold.

[4] Here $P^{(H)}$ is the power set operation in $V^{(H)}$.

algebra of external truth values of $\mathsf{Set}^{(H)}$ is isomorphic to H. If B is a complete Boolean algebra, $\mathsf{Set}^{(B)}$ is a Boolean topos whose algebra of external truth values is isomorphic to B.

There is an alternative description of the topos $\mathsf{Set}^{(H)}$ employing the notion of an *H-set*. This defined to be a set X equipped with an 'H-valued equality relation' $\delta\colon X \times X \to H$ satisfying

$$\delta(x,\ x') = \delta(x',\ x) \qquad \delta(x,\ x') \wedge \delta(x',\ x'') \leq \delta(x,\ x'').$$

We shall write $[\![x = x']\!]$ for $\delta(x,x')$. The category Set_H of H-sets has as objects all H-sets; its arrows are 'H-valued functional relations'. That is, an arrow between two H-sets X and Y is a function $f\colon X \times Y \to H$ satisfying

$$[\![x = x']\!] \wedge f(x,\ y) \leq f(x',\ y) \qquad f(x,\ y) \wedge [\![y = y']\!] \leq f(x,\ y')$$

$$f(x,\ y) \wedge f(x,\ y') \leq [\![y = y']\!]$$

$$\bigvee_{y \in Y} f(x,\ y) = [\![x = x]\!].$$

Given two arrows $(X,\ \delta) \xrightarrow{f} (Y,\ \varepsilon)$ and $(Y,\ \varepsilon) \xrightarrow{g} (Z,\ \eta)$, the composite $(X,\ \delta) \xrightarrow{g \circ f} (Z,\ \eta)$, is defined by

$$(g \circ f)(x,\ z) = \bigvee_{y \in Y} f(x,\ y) \wedge g(y,\ z),$$

while the identity arrow on $(X,\ \delta)$ is just δ.

It can be shown that Set_H and $\mathsf{Set}^{(H)}$ are *equivalent* categories. With each $u \in V^{(H)}$ we associate the H-set $\tilde{u} = (\mathrm{dom}(u),\ \delta_u)$ where $\delta_u(x,\ y) = [\![x \in u \wedge x = y]\!]$; and if u, v and $f \in V^{(H)}$ are such that $V^{(H)} \models f\colon u \to v$, we obtain an arrow $\tilde{f}\colon \tilde{u} \to \tilde{v}$ by defining $\tilde{f}(x,\ y) = [\![f(x) = y]\!]$. This procedure yields a functor from $\mathsf{Set}^{(H)}$ to Set_H.

Now let $(X,\ \delta)$ be an H-set. For each $x \in X$ define $\dot{x} \in V^{(H)}$ by $\mathrm{dom}(\dot{x}) = \{\hat{z}\colon z \in X\}$ and $\dot{x}(\hat{z}) = \delta(x,\ z)$. Define $X^{\dagger} \in V^{(H)}$ by $\mathrm{dom}(X^{\dagger}) = \{\dot{x}\colon x \in X\}$ and $X^{\dagger}(\dot{x}) = \delta(x,\ x)$. $X^{\dagger}$ is taken to correspond to X. Given an arrow $f\colon (X,\ \delta) \to (Y,\varepsilon)$ in Set_H, define $f^{\dagger} \in V^{(H)}$ by $\mathrm{dom}(f^{\dagger}) = \{\langle \dot{x},\ \dot{y}\rangle^{(H)}\colon x \in X, y \in Y\}$ and $f^{\dagger}(\langle \dot{x}, \dot{y}\rangle^{(H)}) = f(x,\ y)$. It is not difficult to verify that $V^{(H)} \models f^{\dagger}\colon X^{\dagger} \to Y^{\dagger}$. This gives us a functor from Set_H to $\mathsf{Set}^{(H)}$.

It can then be verified that these two functors are quasi-inverse and so define an equivalence between $\mathsf{Set}^{(H)}$ and Set_H.

It can also be shown (Higgs 1973, Fourman and Scott 1979) that $\mathsf{Shv}(H)$ and Set_H are equivalent categories. Accordingly all three toposes $\mathsf{Set}^{(H)}$, Set_H, and $\mathsf{Shv}(H)$ are equivalent, yielding two alternative ways of describing (the topos of)

H-valued sets: as sheaves on H, or as sets equipped with an H-valued equality relation.

Finally, we mention that toposes of Heyting-algebra (or Boolean-valued) sets can be characterized in categorical terms. A collection $\mathscr{G}$ of objects of a category is called a *generating class* if for any pair of distinct arrows $f, g\colon A \to B$ there is a member G of $\mathscr{G}$ and an arrow $h\colon G \to A$ such that $f \circ h \neq g \circ h$. A topos is said to be *extensional* if $\{1\}$ is a generating class, and *subextensional* if the collection of subobjects of 1 is a generating class, that is, if for any distinct arrows $f,\ g\colon A \to B$ there is a subobject $U \rightarrowtail 1$ and an arrow $h\colon U \to A$ such that $f \circ h \neq g \circ h$. It can then be shown (see Bell 1988) that *a topos is equivalent to a topos of the form* $\mathsf{Set}^{(H)}$ *if and only if it is subextensional and admits arbitrary set indexed coproducts of* 1. *Moreover, a topos is equivalent to a topos of the form* $\mathsf{Set}^{(B)}$, *for a complete Boolean algebra* B, *if and only if it admits arbitrary coproducts of* 1 *and satisfies the axiom of choice.* For Set itself we have the following characterization: *to be equivalent to* Set *it is necessary and sufficient that it be extensional and admit arbitrary coproducts of* 1, *or that it be bivalent and satisfy the axiom of choice, and admit arbitrary coproducts of* 1.

The role of the axiom of choice in characterizing the classical universe of sets in categorical terms is quite striking.

HISTORICAL NOTES

Chapter 1   Boolean-valued models of set theory were first introduced by Scott and Solovay (see Scott 1967) and Vopěnka (1967). Most of the results of this chapter appear in Scott (1967).

Chapter 2   The method of forcing was invented by Cohen, who also proved the independence of the axiom of constructibility and of the continuum hypothesis: see Cohen (2011, forthcoming). Corollary 2.13 is due to Solovay (1965). The Boolean-valued versions of these results appear in Scott (1967) and Vopěnka (1967). For an approach to forcing closely related to the Boolean-valued one given here, see Shoenfield (1971).

The results in Problems 2.14–2.16 appear in Scott (1967). Problem 2.20 is due to Solovay (1963).

Chapter 3   The independence of the axiom of choice from ZF was established by Cohen in 1963 (see Cohen 1966). The Boolean-valued version may be found in Scott (1967) and Vopěnka (1967). Corollary 3.9. is due to Feferman (1965).

Chapter 4   The construction of generic extensions of models of set theory is due to Cohen (1963, 1964); the approach in this chapter is closely related to that of Shoenfield (1971). Theorems 4.6, 4.15, and 4.19 are from Bell (1976a). The result in Problem 4.27 is due to Mansfield and Dawson (1976), that in Problem 4.33 is due to Bell (1976), Problem 4.36 is due to Solovay (1970), Problem 4.37 is due to Solovay (*v.* also Grigorieff 1975), Problems 4.38 and 4.39 due to Vopěnka (*v.* Grigorieff 1975).

Chapter 5   The concept of a collapsing algebra, Corollary 5.2, and Problems 5.3 and 5.4 are due to Levy. The result in Problem 5.5 is due to Levy and Solovay (1967). Results 5.8–5.12 are due to Ellentuck (1976). Theorem 5.13 is a result of Solovay (1966) and Theorem 5.14 by Kripke (1967). Problem 5.16 is due to Bell (1975).

Chapter 6   The formulation of Souslin's hypothesis in terms of trees is due to Miller (1943). The independence of SH from ZFC is due independently to Jech (1967) and Tennenbaum (1968): I have elected to present the latter's construction. The relative consistency of SH is due to Solovay and Tennenbaum (1971), on which my exposition is largely based. Martin's axiom (which was independently formulated by Rowbottom) makes its first appearance in print in Martin and Solovay (1970), where various applications are presented. Problems 6.35–6.37 are drawn from Solovay and Tennenbaum (1971), and Problem 6.38 from Bell (1983).

Chapter 7 Boolean-valued analysis using measure algebras was introduced by Scott (1969), and later developed by Takeuti. An account of the latter's work in this area appears in Takeuti (1978): this has provided the source of much of my exposition. There also the use of algebras of projections as a basis for Boolean-valued analysis is introduced and extensively developed. The idea of interpreting quantum theory in terms of the correspondence between projection-algebra-valued real numbers and commuting self-adjoint operators is due to Davis (1977). Boolean-valued analysis is further developed in Kusraev and Kutateladze (1999).

Chapter 8 Intuitionistic set theory has emerged at the hands of a number of people: for my exposition I have chosen Grayson (1979). That the axiom of choice intuitionistically implies the law of excluded middle was proved in Goodman and Myhill (1978), reformulating in purely logical terms the fundamental result of Diaconescu (1975). The derivation of that law from the Schröder–Bernstein theorem was carried out in a category-theoretic setting by Banaschewski and Brümmer (1986): the proof given here extends their result to intuitionistic set theory. Banaschewski and Bhutani (1986) showed that the Stone representation theorem implies the law of excluded middle within the context of so-called localic toposes; the extension of their result to intuitionistic set theory, and the proof given here, is due to Bell (1999).

It seems to have been Higgs (1973) who first extended to Heyting algebras the theory of Boolean-valued models, but the systematic development of the idea, and its use in providing models of intuitionistic set theory, is due mainly to Grayson (1975, 1979). In particular, the observation that Zorn's lemma holds in Heyting-algebra-valued models was made by Grayson (1975).

Appendix The concept of elementary topos was invented by Lawvere and Tierney (Lawvere 1971, Tierney 1972) in the late 1960s: it has proved to be an idea of immense fertility. Diaconescu's theorem appears in Diaconescu (1975). The idea of an H-set was introduced and developed by Fourman and Scott (1979), where its connections with sheaf theory are explicitly worked out. H-sets were independently invented and studied by Higgs (1973): there one finds, in particular, the equivalence between $\mathsf{Set}^{(H)}$ and Set_H.

BIBLIOGRAPHY

Balbes, R. and Dwinger, P. (1974). *Distributive Lattices.* University of Missouri Press, Columbia, Missouri.

Banaschewski, B. and Bhutani, L. (1986). Boolean algebras in a localic topos. *Math. Proc. Camb. Philos. Soc.* **100** (1), 43–55.

Banaschewski, B. and Brümmer, G. (1986). Thoughts on the Cantor–Bernstein theorem. *Quaestiones Math.* **9** (1–4), 1–27.

Bell, J. L. (1975). A characterization of complete Boolean algebras. *J. London Math. Soc.* **12** (2), 86–88.

Bell, J. L. (1976). Uncountable Standard models of ZFC+V $\neq$ L. *Set Theory and Hierarchy Theory: A Memorial Tribute to A. Mostowski, Birutowice, Poland 1975.* Lecture Notes in Mathematics, vol. 537, Springer, Berlin–Heidelberg–New York.

Bell, J. L. (1976a). A note on generic ultrafilters. *Z. Math. Logik* **22**, 307–310.

Bell, J. L. (1981). Isomorphism of structures in S-toposes. *J. Symbolic Logic* **46** (3), 449–459.

Bell, J. L. (1983). On the strength of the Sikorski extension theorem for Boolean algebras. *J. Symbolic Logic* **48** (3), 841–846.

Bell, J. L. (1988). *Toposes and Local Set Theories: An Introduction.* Oxford Logic Guides 14. Clarendon Press, Oxford.

Bell, J. L. (1997). Zorn's lemma and complete Boolean algebras in intuitionistic type theories. *J. Symbolic Logic* **62** (4), 1265–1279.

Bell, J. L. (1999). Boolean algebras and distributive lattices treated constructively. *Math. Logic Quart.* **45**, 135–143.

Bell, J. L. and Machover, M. (1977). *A Course in Mathematical Logic.* North-Holland, Amsterdam.

Cohen, P. J. (1963). The independence of the continuum hypothesis I. *Proc. Nat. Acad. Sci. USA* **50**, 1143–1148.

Cohen, P. J. (1964). The independence of the continuum hypothesis II. *Proc. Nat. Acad. Sci. USA* **51**, 105–110.

Cohen, P. J. (1966). *Set Theory and the Continuum Hypothesis.* Benjamin, New York.

Cohen, P. J. (2011, forthcoming) My interaction with Kurt Gödel: the man and his work. In M. Baaz, C. Papadeutrow, H. Putnam, D. Scott and C. L. Harper, Jr., eds. Kurt Gödel and the Foundations of Mathematics: Horizons of Truth, Cambridge University Press.

Davis, M. (1977). A relativity principle in quantum mechanics. *Int. J. Theor. Phys.* **16**, 867–874.

Devlin, K. J. (1977). Constructibility. In J. Barwise, ed., *Handbook of Mathematical Logic*. North-Holland, Amsterdam.

Devlin, K. J. and Johnsbraten, H. (1974). *The Souslin Problem*. Lecture Notes in Mathematics, vol. 405, Springer, Berlin–Heidelberg–New York.

Diaconescu, R. (1975a). Axiom of choice and complementation. *Proc. Amer. Math. Soc.* **51**, 176–178.

Dickmann, M. (1975). *Large Infinitary Languages*. North-Holland, Amsterdam.

Drake, F. R. (1974). *Set Theory: An Introduction to Large Cardinals*. North-Holland, Amsterdam.

Easton, W. B. (1970). Powers of regular cardinals. *Ann. Math. Logic* **1**, 141–178.

Ellentuck, E. (1976). Categoricity regained. *J. Symbolic Logic* **41**, 639–643.

Feferman, S. (1965). Some applications of the notions of forcing and generic sets. *Fund. Math.* **56**, 325–345.

Felgner, U. (1971). *Models of ZF-set Theory*. Lecture Notes in Mathematics, vol. 223, Springer, Berlin–Heidelberg–New York.

Fitting, M. C. (1969). *Intuitionistic Logic, Model Theory and Forcing*. North-Holland, Amsterdam.

Fourman, M. P. and Hyland, J. M. E. (1979). Sheaf models for analysis. In Fourman Mulvey and Scott eds. (1979), pp. 280–302.

Fourman, M. P. and Scott, D. S. (1979). Sheaves and logic. In Fourman Mulvey and Scott eds. (1979), pp. 302–401.

Fourman, M. P., Mulvey, C. J., and Scott, D. S. eds. (1979). *Applications of Sheaves. Proc. L.M.S. Durham Symposium* 1977. Lecture Notes in Mathematics, vol. 753, Springer, Berlin–Heidelberg–New York.

Goldblatt, R. I. (1979). *Topoi: The Categorical Analysis of Logic*. North-Holland, Amsterdam.

Goodman, N. and Myhill, J. (1978). Choice implies excluded middle. *Z. Math. Logik Grundlag. Math.* **24** (5), 461.

Grayson, R. J. (1975). A sheaf approach to models of set theory. M.Sc. thesis, Oxford University.

Grayson, R. J. (1978). Intuitionistic set theory. D. Phil. Thesis, Oxford University.

Grayson, R. J. (1979). Heyting-valued models for intuitionistic set theory. In Fourman Mulvey and Scott eds. (1979), pp. 402–414.

Grigorieff, S. (1975). Intermediate submodels and generic extensions in set theory. *Ann. Math.* **101**, 447–490.

Halmos, P. R. (1963). *Lectures on Boolean Algebras*. Van Nostrand, New York.

Halmos, P. R. (1965). *Measure Theory*. Van Nostrand, New York.

Higgs, D. (1973). A category approach to Boolean-valued set theory. (Unpublished typescript, University of Waterloo.)

Jauch, J. M. (1968). *Foundations of Quantum Mechanics*. Addison-Wesley, Reading, MA.

Jech, T. J. (1967). Non-provability of Souslin's hypothesis. *Comment. Math. Univ. Carolinae* **8**, 291–305.

Jech, T. J. (1971). *Lectures in Set Theory*. Lecture Notes in Mathematics, vol. 217, Springer, Berlin–Heidelberg–New York.

Jech, T. J. (1973). *The Axiom of Choice*. North-Holland, Amsterdam.

Jech, T. J., ed. (1974). Axiomatic Set Theory. *AMS Proceedings of Symosia in Pure Mathematics*, vol. XIII, Part II. American Mathematical Society, Providence.

Johnstone, P. T. (1977). *Topos Theory*. Academic Press, London.

Johnstone, P. T. (1982). *Stone Spaces*. Cambridge Studies in Advanced Mathematics 3. Cambridge University Press, Cambridge.

Johnstone, P. T. (2002). *Sketches of an Elephant: A Topos Theory Compendium*, vols. I and II. Oxford Logic Guides vols. 43 and 44, Clarendon Press, Oxford.

Kelley, J. L. (1955). *General Topology*. Van Nostrand, New York.

Kripke, S. (1967). An extension of a theorem of Gaifman-Hales-Solovay. *Fund. Math.* **61**, 29–32.

Kunen, K. (1980). *Set Theory*. North-Holland, Amsterdam.

Kurraev, A. G. and S. S. Kutateladze (1999). Boolean Valued Analysis. Kluwer, Pordrecht.

Lambek, J. and Scott, P. J. (1986). *Introduction to Higher-Order Categorical Logic*. Cambridge University Press, Cambridge.

Lawvere, F. W. (1971). Quantifiers and sheaves. In *Actes du Congrès Intern. Des. Math. Nice 1970*, tome I. Gauthier-Villars, Paris, pp. 329–334.

Levy, A. (1965). Definability in axiomatic set theory I. *Proceedings of the 1964 International Congress on Logic, Methodology and Philosophy of Science*. North-Holland, Amsterdam.

Levy, A. and Solovay, R. M. (1967). Measurable cardinals and the continuum hypothesis. *Israel J. Math.* **5**, 234–238.

Mac Lane, S. (1971). *Categories for the Working Mathematician*. Springer-Verlag, Berlin.

Mac Lane, S. (1975). Sets, topoi, and internal logic in categories. In H. E. Rose and J. C. Shepherdson, eds., *Logic Colloquium 73*, pp. 119–134. North-Holland, Amsterdam.

Mac Lane, S. and Moerdijk, I. (1992). *Sheaves in Geometry and Logic: A First Introduction to Topos Theory*. Springer-Verlag, Berlin.

Martin, D. A. and Solovay, R. M. (1967). Internal Cohen extensions. *Ann. Math. Logic* **2**, 143–178.

McLarty, C. (1992). *Elementary Categories, Elementary Toposes*. Oxford University Press, Oxford.

Miller, E. W. (1943). A note on Souslin's problem. *Amer. J. Math.* **65**, 673–678.

Mitchell, W. (1972). Boolean topoi and the theory of sets. *J. Pure Appl. Algebra* **2**, 261–274.

Osius, G. (1974a). Categorical Set Theory: a characterization of the category of sets. *J. Pure Appl. Algebra* **4**, 79–119.

Rasiowa, H. and Sikorski, R. (1963). *The Mathematics of Metamathematics.* PWN, Warsaw.

Rosser, J. B. (1969). *Simplified Independence Proofs: Boolean-Valued Models of Set Theory.* Academic Press, New York.

Rudin, M. E. (1977). Martin's Axiom. In J. Barwise, ed., *Handbook of Mathematical Logic.* North-Holland, Amsterdam.

Scott, D. S. (1967). *Boolean-Valued Models for Set Theory.* Mimeographed notes for the 1967 American Math. Soc. Symposium on axiomatic set theory.

Scott, D. S. (1968). Extending the topological interpretation to intuitionistic analysis I. *Compositio Math.* **20**, 194–210.

Scott, D. S. (1969). Boolean-valued models and non-standard analysis. *Applications of Model Theory to Analysis, Algebra and Probability.* Holt, Reinhart and Winston, New York.

Scott, D. S. (1970). Extending the topological interpretation to intuitionistic analysis II. In A. Kino, J. Myhill, and R. E. Vesley, eds., *Intuitionism and Proof Theory*, pp. 235–256. North-Holland, Amsterdam.

Scott, D. S., ed. (1971). Axiomatic Set Theory. *AMS Proceedings of Symosia in Pure Mathematics*, vol. XIII, Part I. American Mathematical Society, Providence.

Shoenfield, J. R. (1971). Unramified forcing. In Scott, ed. (1971).

Sikorski, R. (1964). *Boolean Algebras.* Springer, Berlin–Heidelberg–New York.

Solovay, R. M. (1965). $2^{\aleph_0}$ can be anything it ought to be. In J. W. Addison, L. Henkin, and A. Tarski, eds., *The Theory of Models.* North-Holland, Amsterdam.

Solovay, R. M. (1966). New proof of a theorem of Gaifman and Hales. *Bull. Amer. Math. Soc.* **72**, 282–284.

Solovay, R. M. (1970). A model of set theory in which every set of reals is Lebesgue measurable. *Ann. Math.* **92**, 1–56.

Solovay, R. M. and Tennenbaum, S. (1971). Iterated Cohen extensions and Souslin's problem. *Ann. Math.* **94**, 201–245.

Takeuti, G. (1978). *Two Applications of Logic to Mathematics.* Princeton University Press, Princeton.

Takeuti, G. and Zaring, W. M. (1973). *Axiomatic Set Theory.* Springer, Berlin–Heidelberg–New York.

Tennenbaum, S. (1968). Souslin's problem. *Proc. Nat. Acad. Sci. USA* **59**, 60–63.

Tierney, M. (1972). Sheaf theory and the continuum hypothesis. In F. W. Lawvere, ed., *Toposes, Algebraic Geometry and Logic.* Springer Lecture Notes in Math. 274, pp. 13–42.

Vopěnka, P. (1965). The limits of sheaves and applications on constructions of models. *Bull. Acad. Polon. Sci. Ser. Sci. Math. Astron. Phys.* **13**, 189–192.

Vopěnka, P. (1967). General theory of ∇-models. *Comment. Math. Univ. Carolinae* **8**, 145–170.

Vopěnka, P. and Hajek, P. (1972). *The Theory of Semisets.* North-Holland, Amsterdam.

INDEX OF SYMBOLS

INDEX